THE BEAUTY
OF NUMBERS
IN NATURE

THE BEAUTY

Mathematical patterns and principles

OF NUMBERS

from the natural world

IN NATURE

IAN STEWART

IVY PRESS

This edition first published in the UK in 2017 by
Ivy Press
Ovest House
58 West Street
Brighton
BN1 2RA
United Kingdom
www.quartoknows.com

First published in 2001

Text copyright © Joat Enterprises 2001, 2017

Design and layout copyright © The Ivy Press Limited 2001, 2017

British Library Cataloguing-in-Publication Data

A catalogue record for this book is available from the British Library

ISBN: 978-1-78240-471-2

This book was conceived, designed and produced by
Ivy Press
Publisher Susan Kelly
Creative Director Michael Whitehead
Editorial Director Steve Luck
Project Editor Fleur Jones
Designer Ginny Zeal
Digital illustrations Lance Cummings
Picture Researcher Katie Greenwood

Printed in China

10 9 8 7 6 5 4 3 2 1

CONTENTS

PREFACE

When I was six, a friend showed me some curious little five-pointed stars that he had found on the beach. They were segments of the stem of a fossilized sea lily. I spent weeks trying to find some more, without success. But I did find some beautiful spiral ammonites. Stars, spirals ... I became aware of a deep mystery: why does nature produce so many patterns?

My early experiences with mathematics, on the other hand, were more prosaic. It all seemed to be about numbers; even algebra is just the use of symbols to represent unknown numbers. If anyone had told me that there are deep connections between mathematics and the elegant geometric shapes of fossils, it would have been a big surprise.

Most children find numbers fascinating at first. But for many, the fascination wears off, battered by years spent doing calculations that often seem meaningless. Unlike most of us, I kept my interest in mathematics. And slowly I discovered two things: that numbers can be fascinating in their own right, and that they are just the tiny tip of a gigantic mathematical iceberg, including many other things: shapes, chance, movement, and above all *patterns*. In fact, mathematics is often described as a systematic theory of patterns.

The patterns of mathematics can be metaphorical. Every square number ends in 0, 1, 4, 5, 6, or 9, but not 2, 3, 7, or 8. That's a pattern, in a sense, but you wouldn't put it on the wall to look pretty. Patterns can also be highly visual— ammonites, snails, whirlpools, and galaxies are all spirals. A honeycomb inside a beehive consists of hundreds of tiny hexagonal cells. The same regular arrangement also arises when packing lots of identical coins together as tightly as possible. That's surprising, because coins are circles, not hexagons. Another occurrence is in the atoms of a crystal of ice, which is why snowflakes are often six-sided. Patterns can also be dynamic, such as regularities of movement. For instance, there's a deep mathematical unity among all forms of animal motion, from slithering snakes to trotting horses.

These examples point to a deep truth: mathematical patterns are universal, with the same pattern appearing in many different contexts.

That's what this book is all about. How mathematics ("numbers") provides important insights into the world we live in ("nature"). Nature's patterns and the mathematics that explain them have a deep appeal to our aesthetic sense ("beauty"). Nature's beauty is direct and visual, whereas the beauty of mathematics is inherent in its logical structure and the insight it provides. However, thanks to today's computer graphics, mathematics can have visual beauty, too.

The mathematics of patterns came from a vast range of sources. The ancient Greek cult of the Pythagoreans and their obsession with numbers as the philosophical basis for the universe; a problem about rabbits in an arithmetic text published in 1202; a great mathematician wondering how a violin makes music; a clerk in a patent office who realized that space and time can get mixed together; and a maverick mathematician who wondered why nature seldom uses regular geometric shapes like spheres and cylinders, preferring jagged lightning bolts, irregularly branching trees, and the tumbled terrain of mountainsides.

This book begins with one simple, typical question: why do so many snowflakes have symmetric, six-sided shapes, yet of such diversity that each is a one-off? How can this strange mixture of regularity and irregularity coexist in what, basically, is a tiny lump of frozen water? By the end of the book we will have an answer, of sorts, which shows that mathematical "patterns" can come in many guises, not all of which look much like patterns at all. There can be patterns in the rules that nature obeys, even when the behavior itself seems pattern-less.

Along the way, the quest to understand the humble snowflake opens up into a broad discussion of the relationship between mathematics and the natural world. Numbers play their part, and so do regular shapes such as hexagons. A deeper element is the idea of structural form, especially symmetry. Innumerable patterns in nature have the same explanation: the underlying physical laws are symmetric, and some—though not necessarily all—of those symmetries appear in the pattern. For example, parallel lines of sand dunes in the desert and the stripes on a tiger both arise from the same symmetry-breaking process: one in sand, the other in chemical pigments.

Another part of the puzzle is dynamics: how things move, how their shapes, sizes, and positions change over time. Isaac Newton discovered, using mathematics invented for the purpose, that the unruly movements of the planets of the solar system can be tamed by understanding one simple, elegant mathematical rule: the law of gravity. Mathematics tells us where to look for explanations of nature's patterns: not in the patterns themselves, but in the underlying laws. Chaos theory tells us how regular laws can sometimes lead to irregular behavior.

All of today's science and technology builds on that insight. Nature obeys rules, which we discover and express using mathematics. The six-fold symmetries of snowflakes, and their diversity of forms that arise, result from simple laws of chemistry and dynamics. There are some who feel that discovering the rules somehow spoils the beauty—much as knowing how a stage magician produces a rabbit from a hat can spoil the illusion. But nature's patterns aren't stage tricks, and understanding their origins reveals new dimensions and relationships that can only add to their beauty.

Ian Stewart

PART ONE

PRINCIPLES & PATTERNS

1

THE PUZZLE

What shape is a snowflake? There it lies, on the sleeve of my coat, sparkling in the light of a streetlight. Snow drifts gently down, small crisp flakes, and it's cold. This is all to the good, since my captured snowflake isn't going to melt before I get to look at it, but my ears are starting to go numb.

Even with the naked eye, I can see that my snowflake isn't just a random lump. It has a definite form. When magnified by my pocket lens—bought specially for the purpose—the sight is breathtaking. My snowflake is like a fern, made of clear crystal. More precisely, it looks like six ferns, all joined at the root and all identical. My snowflake is a puzzling mixture of regularity and randomness, of order and disorder, of pattern and mindless jumble. It has almost perfect sixfold symmetry, six copies of the same

shape—but that shape is like nothing ever seen in Euclid's geometry. It isn't exactly random, but you won't find its name in any dictionary.

You will find the word "dendrite," which is the term scientists use to describe this sort of shape, but it refers to a category of shapes, not any particular one. It comes from the Greek word meaning "tree." What shape is a tree? It is tree-shaped. A snowflake isn't a tree, or a fern, or a feather.

It's a snowflake, and it's snowflake-shaped.

Next to it, another snowflake achieves touchdown. It, too, has sixfold symmetry. It is equally, enigmatically, not quite fernlike. And it is different from the first snowflake. It seems that "snowflake-shaped" begs a crucial question. They say that no two snowflakes are alike, but the mathematician in me can see that this statement is either trivial or a gross exaggeration. Any two objects in the universe are different if you look closely enough—well, maybe not two electrons, though even there I

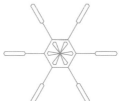

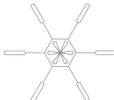

LEFT Seen from a distance, snow has a stark, majestic beauty. Viewed close up, a single flake of snow is a tiny geometric jewel, a vital clue to the intricacy and beauty of nature's patterns.

wonder. But if you accept only those differences that are visible under a low-powered lens, and take into account just how many snowflakes have fallen during the Earth's four billion year history, surely somewhere, some time, some snowflake's double must have appeared? Though if you do the calculations, maybe not. If my eye can distinguish a hundred tiny features, and each can be present or not, that makes a nonillion—a million trillion trillion—different snowflake shapes. At any rate, there's enough diversity in the designs used in snowflake manufacture that I'm not going to find two identical twins tonight.

Like most people, I've known about the shape of snowflakes ever since I was old enough to read encyclopedias. But until now I've just looked at the pictures in the books and occasionally I've given a real snowflake a quick glance with the naked eye. This is the first time I've gone out with a lens and really looked, and, amazingly, they're just like the encyclopedia said—kind of frondy, with a visible hint of hexagon; the mathematician's archetypal six-sided shape. Some snowflakes are just hexagons, with six straight sides, so these ones are all the same shape. I guess they weren't included in the traditional slogan about snowflake diversity, or else it was poetic license. The rest, though, are some kind of second cousin of the hexagon, several times removed, and those are the ones I'm interested in.

What sort of universe makes shapes like snowflakes? It's very mysterious. But as I stand and freeze, one thing is crystal clear. Ice. It has something to do with ice. Snowflakes are made of ice, so it must have something to do with ice.

The ice in my freezer comes in cubes. Well, they're called cubes, but they're only roughly box-shaped. No hexagons there, and, more importantly, no feathery fronds. Anyway, this kind of ice is made in a mold: buy the right mold, and you could have a freezer full of ice teddy bears—or hexagons, but getting them this way is cheating. I doubt there are snowflake molds up in the clouds. Up there, the ice puts itself together into patterns without human intervention. But whatever's going on up there, it definitely has something to do with ice.

JACK FROST

One of my earliest memories is of ice on the inside of my bedroom window—feathery, leafy patterns of ice, Jack Frost.

We lived in an end-of-terrace house with bay windows. In winter, the room was heated by a coal fire. Overnight, the fire would die down and go out. Water from the air would condense on the cold windows and freeze as the night wore on. And in the morning, there it was—all ferns and leaves like a surreal jungle in a Rousseau painting. Of course I didn't think about it that way then, but it was pretty and very puzzling.

RIGHT Water is not just two atoms of hydrogen to one of oxygen. It is more like a room full of whirling dancers. As water freezes into ice, the dancers stop— but their frenzied exertions determine where they come to rest.

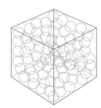  

It's not something you often see on windows in these days of central heating. But leave a car parked outdoors on a frosty night, and next morning you'll find the same patterns all over the windshield, and possibly over the whole car. Apparently, ice likes to make these fernlike patterns all by itself. It just needs something cold to grow on. There aren't any windows up in the clouds, though. And Jack Frost on a windowpane isn't six-sided. It's just an irregular leafy jungle. Still, these ice fronds look like a good start. I wonder what causes them?

Ice makes many shapes. Even up in the clouds, it also makes needles and tubes and (rarely) triangles, not to mention pyramids and bullets and capped columns. And hail. Boy, does it make hail. A few years ago, in Minneapolis, we saw the mid-afternoon skies darken like sudden midnight. Hail the size of golf balls pounded down from the heavens, hammering dents into cars and causing pedestrians to flee for their lives. The hailstones' weight, aided by high winds, uprooted huge trees. That was in the spring. During the long Minnesota winter, ice and its innumerable physical forms are inescapable. Snowdrifts, icicles…Minnesota is mostly lakes, so in winter it is mostly ice.

What is ice? Frozen water. What is water? Chemistry textbooks tell us that it is a simple molecule, two atoms of hydrogen to one of

LEFT When crystals of ice grow on a cold surface, they do not spread evenly across it, they form jumbled, glittering forests of jagged ferns. There are at least sixteen forms of ice: the snowflake will not give up its secrets easily.

H_2O

oxygen. This simplicity is deceptive, for water is one of the subtlest and most puzzling of liquids. It dissolves an astonishing range of chemicals, and it is a key ingredient of our form of life.

I like to think that out there in the vast universe, there are other forms of life, quite different in substance from our own, but resembling earthly life in its deeper features of complexity and self-organization. Alien life forms might not encode their genetics in DNA, might not be made of carbon, might not require water—or a planet, or an atmosphere. They might not even be made of matter: imagine, for example a creature made from interwoven magnetic vortices in the surface of a star. But until we meet such creatures—and we could be alone—the life-forms that we know rely heavily on the strange properties of water.

This is why NASA has become interested in Jupiter's moon Europa, which on the face of it doesn't look like the sort of place to harbor life, since its surface is half a mile-thick layer of ice. However, there's clear evidence that beneath the ice there is an ocean of liquid water more than 60 miles deep. So there's more water in Europa's ocean than in all of the Earth's oceans combined. Tidal forces from Jupiter knead Europa's core and keep it warm, so there's also an energy source. Some 2 miles (4km) beneath the Antarctic ice one of Earth's biggest lakes, Lake Vostok, lies buried. We know there are bacteria in Lake Vostok, so why not beneath Europa's surface too?

Water is an unusual chemical. It comes as a gas (steam) and a solid (ice) as well as a liquid. Only the gaseous state seems straightforward.

THE MYSTIC

What shape is a snowflake? I can imagine a distant ancestor, a sharp-eyed protohuman, puzzling over the tiny white specks deposited on its hairy coat. I know that the sixfold symmetry

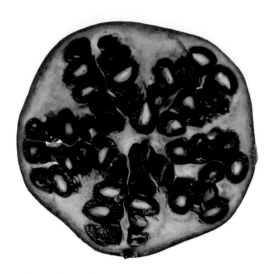

of snowflakes has been remarked on for thousands of years. And among my books is a much-thumbed copy of *On the Six-Cornered Snowflake*, written in 1611 by the German astronomer Johannes Kepler as a New Year's gift for his sponsor, John Matthew Wacker of Wackenfels. Kepler was an inveterate pattern hunter. He saw mathematical laws in the seeds of pomegranates and the movements of the planets, and some of these laws still inform today's scientific thought. Pomegranate seeds embody important features of the three-dimensional geometry of close-packed units: in the pomegranate, evolution has homed in on an efficient way to pack as many seeds as possible into a confined space. And the movements of the planets led Kepler to describe a highly accurate form for their orbits—ellipses. Between 1609 and 1619 he worked out how their periods of revolution around the Sun, and their speeds in orbit, depend on their distance from that central star. Fifty years later, his discoveries led Isaac Newton to formulate his law of gravity, which survived unchanged and unchallenged for 250 years and was only bettered by Albert

FAR LEFT Nature's patterns hint at the underlying regularities of physics. The way seeds are arranged in a pomegranate led Kepler to an understanding of the structure of ice.

LEFT The stars in the sky look random, but even here there is a pattern—the way they move. Every night, the stars appear to revolve in circular arcs centered on the Pole Star. This geometry tells us that it is not the sky that rotates, but the Earth. As our planet spins, the stars seem to do the same. The Pole Star seems to stay fixed because it is aligned with the Earth's axis.

Einstein. Newton's law is still accurate enough to put humans on the Moon. Kepler also saw— or thought he saw—a mathematical law in the spacings of the planets, but this law has failed the test of time. For a start, it required there to be only six planets and we now recognize eight.

Despite a few misses, Kepler's scientific strike-rate was remarkable. Nearly 400 years ago, when physics was just a gleam in Galileo's eye, this mathematician-cum-mystic posed the riddle of the snowflake: "There must be some definite cause why, whenever snow begins to fall, its initial formations invariably display the shape of a six-cornered starlet. For if it happens by chance, why do they not fall just as well with five corners or with seven? Why always with six, so long as they are not tumbled and tangled in masses by irregular drifting, but still remain widespread and scattered?" By drawing on his extensive experience of nature's patterns and their mathematical analogs, Kepler came up with a pretty good explanation of the snowflake's sixfold symmetry. The fronds, the countless variants—these were too much even for him. But the basic sixfold character of a snowflake, that he could handle.

I want to go further than Kepler did. I want to understand all the patterns of nature—not just snowflakes. I don't expect to succeed, but I want to get as far as I can. There are many other puzzles in the natural world: the spiral of a snail shell; the waves that run along the legs of a moving millipede; the serried cells of a honeycomb or a wasp's nest; the multicolored arc of a rainbow; the stripes of a tiger; the jagged slopes of a mountain range; the blue-white sphere of the Earth seen from space; the celestial river of the Milky Way, 400 billion stars of which our Sun is but one; the ghostly vortex of the Andromeda galaxy; the form of the universe itself, and the bizarre physics of the particles from which it is made.

Where do nature's patterns come from? What makes them? Kepler could not even have dreamt of some of the subject matter, but he would have liked the questions—and he would have been amazed at how closely their answers are connected to his beloved snowflake. I'm going on a journey in search of the snowflake's secret…and, with it, the deeper secrets of our astonishing universe. And you're coming with me.

2

NATURE'S PATTERNS

Mathematicians seek generalities. They're unimpressed by one triangle whose angles add up to 180 degrees, but they find it striking that the same is true of any triangle. When seeking generalities, it helps to have a lot of examples because then you can compare and contrast and try to extract the essence of whatever it is you are seeking. A snowflake is one example of pattern formation in nature; but the secret of the snowflake is unlikely to emerge from concealment without a wide-ranging search.

Earth is the right place to be when looking for patterns. Where the rest of the solar system has rocks (hot or cold but hardly ever "just right"), planet Earth has plants and animals. Both organic and inorganic worlds exhibit striking patterns—butterflies and rainbows show the same range of colors, for instance—but plants and animals display an apparently effortless riot of color and form that the inorganic world struggles to match.

One of the commonest patterns in animal markings is stripes. Sometimes the stripes are very regular, the kind of thing you would associate with mathematics—parallel lines of alternating colors, black and white, purple and yellow. More or less regular stripes are common in tropical fish—for that matter, almost any pattern is common in tropical fish—and in seashells. The emperor angelfish deserves its name for its imperial dress: brilliant gold and purple, tinged with ermine, and dominated by narrow black-and-white stripes that run most of the length of its body. They're not completely regular, sometimes they fork like a Y or glitch sideways, but the regularities dominate. Stripes on seashells come in two kinds: mostly they run either in the same direction as the spirals of the shell or at right angles to them. Other animals such as the raccoon have rings of bold stripes around their tails.

RIGHT One of nature's favorite patterns is the stripe, like those seen on the zebra. Put simply, stripes repeat roughly the same patterns in a series of equally spaced parallel lines.

Of course, when speaking of striped animals it is the zebra and the tiger that spring most readily to mind. The zebra's stripes are brash and bold, by no means perfectly parallel, distinctly more subtle than you might expect from anything mathematical. The stripes do funny things where the legs or tail merge into the main part of the animal's body, and each of the three main species—Burchell's zebra, Grevy's zebra, and the mountain zebra—has its own distinct pattern of stripes. The tiger's stripes are more subtle still, arrayed along the big cat's flanks like brush-strokes, an elegant exercise in living calligraphy. William Blake memorably wrote of the tiger's "fearful symmetry," probably referring to its elegant, powerful form; but to a mathematician, the tiger's symmetry includes its stripes, and not just metaphorically.

The inorganic world has its own stripes, too. Waves rolling up a beach are typically arranged in long, parallel rows—peak and trough in place of black and white. In the deep desert, the simplest patterns of sand dunes are transverse dunes, stripes of sand piled at right angles to the prevailing winds, and linear dunes, stripes arranged at an angle to more variable winds. You also find stripes in rocks. In Australia there is the world's only deposit of zebra rock, which looks like striped candy (and is unfortunately about to be submerged by a reservoir). But the stripes in rocks are a record of their history, as they were deposited layer by layer on the bed of an estuary or a shallow sea. In contrast, the stripes of waves and dunes are created on a far less leisurely timescale. Ocean waves, especially, reflect the dynamics of the present.

Stripes in fur, stripes in flesh, stripes in sand, stripes in water—are they as different as they seem? Is the similarity between them no more than a visual pun? Or is there a hidden unity? Is there a universal mechanism that creates stripes? If my search for the secret of the snowflake is to succeed, there has to be a big picture into which a variety of nature's patterns fit. If I can't find unity in some portion of the world's stripes, where will I be able to find it?

LEFT AND ABOVE There are stripes on fish and even stripes in the ocean, which we call waves. Does the geometric similarity between different instances of stripes indicate a hidden mathematical unity, or is it just coincidence?

SCULPTED IN SAND

Stripes are often found in association with more complex patterns, for example wave markings left in the sand at low tide, or sand dunes in the desert. Deserts, with their striking structure, are a laboratory of pattern formation, although we actually know remarkably little about how patterns form in sand.

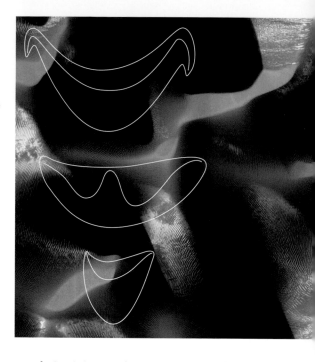

The physics of dunes is deceptively simple. As the wind blows across the desert, it picks up grains of sand from some places and deposits them in others. The shape of existing dunes affects the flow of air, thereby affecting how much sand gets deposited or eroded, and where. So there is a dynamic coevolution of the patterns of air flow and the patterns of heaped sand.

The results of this coevolution are fascinating, and far from obvious. Simple patterns of parallel stripes are only the beginning. Big dunes are covered in small corrugations. The parallel ridges of transverse dunes can themselves become wavy, a pattern of repeated zigzags known as barchanoid ridges. The lines of heaped sand can break up altogether, forming crescent-shaped barchan dunes with the horns of the crescents trailing away from the wind. Alternatively, they can form parabolic dunes, crescent-shaped scoops of sand culminating in curved crests, whose horns point into the wind. If the wind's direction is variable, the desert can arrange itself in an array of dome dunes, each dome a smooth circular mound. If the wind's direction is very variable, these domes can turn into stars, speckling the desert for many hundreds of miles.

Barchan sand dune patterns have also been seen on Mars. Saturn's moon Titan has the largest known field of linear dunes in the solar system.

Our theoretical understanding of the myriad patterns of the desert rests mainly on computer simulations. The physics may sound simple, but translating it into mathematics is extremely hard. Sand is composed of individual grains, unlike the finely divisible fluids of air or water, so one of the mathematician's favorite tricks—creating an idealized model of an infinitely divisible continuum—doesn't work particularly well for sand. Dune geometry is also a multiphase flow problem—it involves air as well as sand, one a genuine fluid, the other a somewhat granular one. The boundary between sand and air is part of the answer, not part of the assumptions, and it changes as time passes. And when a sandstorm surges over the desert, one of the main things that determines the shapes and positions of dunes, the boundary between the sand and the air, becomes too fuzzy to specify.

It all looks pretty hopeless, and it is if you insist on tackling the problem directly. But, as I've already said, mathematicians like to think in generalities. Maybe there are other, more tractable systems that exhibit similar patterns to those that can be found in the desert. Maybe ocean waves can tell us something about transverse dunes. Maybe dunes can tell us about

the zebra. Maybe something quite different from any of these can unlock the common secret of all the world's stripes.

One fruitful line of attack is to ask why patterns form at all. Much the same wind conditions exist across huge swathes of desert, and even if the wind varies, it varies in much the same manner everywhere. As the wind sweeps across the desert, then, why doesn't it smooth the sand out and lay it flat, like a knife smoothing out the icing on a cake? By the same token, why are oceans always a mass of heaving waves? Why does the water slop around so much? And when it does slop around, how come it slops in recognizable patterns, rather than slopping in a random muddle? Already there's a common unity of questions, and we haven't even started to look for answers to them yet. To press home the point: since the entire body of a tiger or zebra is covered in hairs, which have much the same structure everywhere on the animals' bodies, why does the pigment in the hairs create colored regions with visible regularities? Why aren't zebras gray all over?

ABOVE Even using materials as simple as grains of sand, nature can sculpt elegant patterns. From the chaos of whirling windborne sand emerges the order of dunes—giant waves of sand, sweeping across the desert, seemingly frozen into hills and valleys, but actually marching across, rank upon rank. Desert sand creates a great variety of patterns, both simple and complex. Sand dunes are like slow ocean waves— and, like water waves, they form tiny ripples as well as huge crests and troughs, like these sand dunes on Mars. The physics of sand is mysterious, but some general pattern features are starting to be understood.

BELOW AND RIGHT Honeycomb patterns are less ubiquitous than stripes, but still quite common in the natural world. Their underlying mathematical structure is more evident than that of stripes, because honeycombs are built from hexagons, and hexagons come straight out of the geometry texts. Honeycombs also exemplify a key feature of pattern formation: the use of the same basic unit, repeated over and over again. A honeycomb is an efficient way to fit together a lot of identical, roughly circular regions, and nature has many reasons for doing just that—so the hexagon and its associated honeycomb play a key role in pattern formation, in both physics and biology.

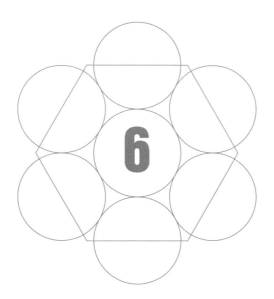

HONEYCOMBS

There's more to life's patterns than stripes. Spots, for a start. Why are tigers stripy but leopards spotty? Maybe animal markings can be pretty much anything they want to be. After all, look at the bizarre plumage of birds of paradise. Presumably an animal or bird's genes can instruct its cells to form pretty much any pattern. The curious thing, though, is that on the whole they don't. Most patterns come out of a standard catalog of simple forms—stripes, spots, uniform areas of color. Often these patterns are assembled into strange combinations, especially in birds and fish, so maybe the genes put the patterns together into a whole, but something else determines the catalog.

What else is in the catalog? Honeycombs. A honeycomb pattern really does look like mathematics—line upon line of perfect hexagons stacked in a neat two-dimensional array. Honeycombs and snowflakes share the magic number six, a coincidence that was not lost on Kepler. In a beehive the hexagons are tiny chambers, each capable of holding a grub or some honey. I still remember my amazement when the man who deals with our regular wasp infestations produced an old wasp's nest from inside the roof of our house and, with some trepidation, I broke open this miracle of paper engineering. The interior was a beautiful construction of nearly identical hexagonal chambers, though put together in a rather irregular fashion, which suggested that the wasps started building various bits of it at the same time without recourse to any master plan and kind of fudged the joins when the different parts met up.

Bees generally build their honeycombs in the vertical direction, with the hexagonal tunnels running horizontally; wasps build them in the horizontal direction, with the hexagonal tunnels running vertically. How can bees and wasps be smart enough to make such things? They are social insects, somehow capable of much more when they act as a collective than any individual could possibly achieve. I don't think they are that smart. I think that something gives them a helping hand, a head start. Some patterns are easier to make than others, they arise because at heart the universe obeys simple rules.

In consequence, it's no surprise that wasps and bees aren't the only creatures that make honeycombs. So do territorial fish. And here, the reasons for the honeycomb structure are more evident. A typical case is a species of tiny fish in Lake Huron with strong territorial instincts, which carve themselves dish-shaped territories about 12in (30cm) across. Each fish stations itself in the middle and repels all invaders. There are a lot of these fish and their territories are packed closely together. As it happens, they pack in a honeycomb pattern. At first sight, that looks like an amazing engineering feat, but there's a trick. The clue is "packed closely." If you take a lot of identical circles—coins, say—lay them on a table, and jiggle them around until they're squashed as tightly as possible, then they form themselves into a honeycomb. In practice it's not perfectly regular, but the same goes for fish territories and bees' honeycombs. But it's pretty close. About a hundred years ago mathematicians proved that the honeycomb is the most efficient way to pack circles in the plane. The main reason is that six circles fit exactly around another circle of the same size, and the honeycomb repeats this structure around every circle.

At any rate, what this all tells me is that regular large-scale patterns can be created by obeying simple, local rules. This hints at a rationale behind nature's pattern book. Kepler realized this long ago—the regularity of close-packed circles is a key observation in his

book about snowflakes. Fish pack their territories together; bees and wasps have evolved a structure that packs their grubs together.

What does a snowflake pack together?

COAGULATED GLOBULES

Kepler asked himself the same question. He knew that snow was made of condensed water vapor, and he wondered whether it condensed in a definite pattern. "Granted then that vapor coagulates into globules of a definite size, as soon as it begins to feel the onset of cold… Secondly, granted that these balls of vapor have a certain pattern of contact…" This set him off on a wild-goose chase that was based on the mistaken belief that the snowflake is three-dimensional rather than the reality of it basically being flat, but he soon reverted to the geometry of the plane and turned his error into a triumph: "Six corneredness is chosen by the formative faculty from material necessity as well, so that no gap should be left and the

gathering of vapor into formations of snow should take place more smoothly."

His final paragraph related the geometric regularity of snowflakes to the regular geometry of crystals: "It is probable, therefore, that our formative faculty varies with variations in the liquid. In sulfates of metals the rhomboid cubic shape is common, and saltpeter has its own shape. So let the chemists tell us whether there is any salt in a snowflake and what kind of salt, and what shape it assumes otherwise!" Kepler's quest for the secret of the snowflake led him to the broader question of crystal structure, and we must follow the same path.

It might seem evident that the form of crystals shows regular mathematical patterns— for example a crystal of common salt is a cube. However, the idea that crystals are mathematical used to be controversial. Instead of worrying about the regularities of crystal form, people worried about whether those regularities were real or illusory. The word "crystallographer" acquired overtones nowadays associated with

"astrologer" or "ufologist." The French naturalist Count Buffon said in the early 18th century that "all the work of the crystallographers serves only to demonstrate that there is only variety everywhere where they suppose uniformity."

This skepticism was more reasonable than it might seem now. Specimens of crystal found in nature are usually much less regular than those grown in the laboratory. Progress on crystals was slow until, later in the century, the German geologist Abraham Werner invented a classification system for minerals that he called oryctognosy, which outlined how to tell which mineral you've got by observing its color, hardness, density, and so on. Once mineralogists could be sure that two apparently different specimens of a mineral were actually the same mineral—or not—it became possible to look for regularities. These soon became apparent in the angles between the flat facets of crystals. All crystals of a given mineral, however damaged or irregular, exhibit the same characteristic list of angles. Not only that—the same list often shows up in other minerals. Scientists could measure angles, obtain numbers, and seek underlying causes for their patterns. The pattern of the snowflake is all about angles: with angles of 60 degrees and 120 degrees all over the place. Why?

Pattern-seeking mathematicians had already started to explore the new territory, even before it was certain that it existed. Kepler was soon

BELOW Crystals have a regular geometry, which reflects their regular atomic structure. They are built from identical units, in patterns that repeat along three spatial axes.

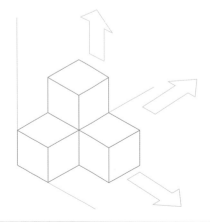

followed by the English scientist Robert Hooke, whose *Micrographia* of 1665 included pictures showing how collections of close-packed circles and spheres could mimic crystalline forms. Around a century later, René Just Haüy, a priest and amateur mineralogist, noticed that when a crystal of calcite splits, it always breaks up into pieces like lopsided boxes, and he suggested replacing Kepler and Hooke's spheres by pieces of this general shape. Crystallographers made strenuous efforts to discover the nature of the basic building blocks of crystals, but ran into difficulties because these building blocks seemed to be vanishingly small.

The impasse was broken by the mathematicians, who decided not to worry about what these building blocks were, but instead asked how they were arranged. It turned out that they were arranged in regular lattices—spatial patterns in which the same basic unit is repeated in three different directions. For a simple example think of filling space with cubes, packed in the obvious way—here the three directions are north, east, and up. This approach led to a classification of the possible symmetries of crystals, which much later led to a solution of the problem that had been worrying the crystallographers all along—what is the nature of the basic building blocks of crystals? It turned out that they were atoms, particles of matter so tiny that until very recently they were invisible under even the most powerful microscope. It was a remarkable case of pure mathematics leading to a significant breakthrough in physics.

SPIRAL SWIRLS

The ancients were confused by the irregularities of natural crystals, which were often damaged or imperfectly shaped because they had grown under variable conditions. Laboratory crystals typically look like flat-faced geometric polyhedrons. Such patterns are, on the whole, a bit too sharp-edged to turn up in living creatures. One of the favorite patterns of life, in fact, is based on curves—the spiral.

We use the word "spiral" in at least two senses: for a flat curve that swirls around and around, moving ever outward as it does so; or for a twisted curve in space like a spiral staircase. Nature uses both of these forms, including a mixture of the two, and nowhere is this more apparent than in shells.

Spiral shells appear way back in the fossil record, and are one of the most common. The flat spiral of the ammonite—of which there have been many distinct species—is widely recognized. As a child, I used to hunt ammonites on the seashore and on most days I found tiny coiled ammonites among the rocks, washed out of the clay by the action of the waves. The local museum had huge ones, a yard or more across, but of course I never found such large specimens.

The shape of some ammonites is close to what mathematicians call an Archimedean spiral—successive coils are spaced the same distance apart. The majority of ammonites, though, form a logarithmic spiral, whose spacing multiplies by a fixed amount for each turn of the spiral. The best known modern shell with this spiral form is the Nautilus, with a shape that is astonishingly regular, divided into successive compartments. The Nautilus offers a clue to the spiral shape of such shells, and to the logarithmic spiral in particular.

The shell is formed by a soft-bodied organism, for protection. New shell material is excreted onto the edge of the existing shell. As the creature's size increases with age, it outgrows its existing chamber and builds an extension onto its house. The rate of growth of the organism affects the spacing of successive coils of the spiral. At one extreme, if the organism doesn't actually grow much, but builds the extension

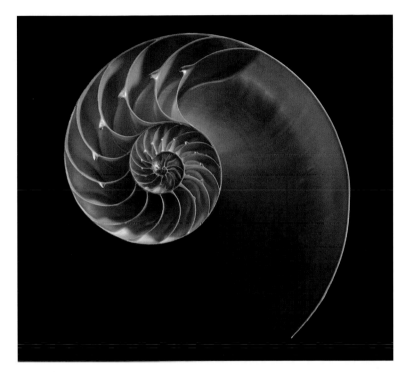

LEFT The shell of the Nautilus, a sea-going mollusk, is made from a series of curved cells, which wind around and around, gradually increasing in size, to form a perfect logarithmic spiral. This mathematical pattern is a clue to how the creature makes its shell, and to how it grows.

anyway—perhaps to use up excreted minerals, though I'm guessing here—then it makes an Archimedean spiral. If instead it grows exponentially, doubling in size in some fixed period of time, it makes a logarithmic spiral. So the flat spirals of ammonites and Nautilus probably result from simple patterns of growth in the creature that builds the shell around itself.

On land, snails build similar shells. Sometimes snail shells coil clockwise, sometimes they coil the other way, and in some species—the exception rather than the rule, which is curious—both directions of twist occur. In 1930 a group of biologists carried out breeding experiments on a million snails and they worked out what determines the direction of coiling. The cause is genetic, but it is not in the snail's genes, it is in its mother's. Amazingly there is a gene for how a snail's children's shells coil, not

for how the individual's shell coils. This ensures that a particular handedness of twist is put into the snail embryo when it reaches a stage of development of just eight cells.

Here's another puzzle to add to the list. Snail shells, and indeed many seashells, often coil into the third dimension. Of course, the shape of the shell is always three-dimensional: what I mean is that the "core" of the spiral, the line that runs along the center of the chambers, ceases to lie in a plane and starts to curl into a third dimension of space. Shells like this are also seen in fossils. The cone-shaped gastropod Turritella is over 50 million years old, dating from the Eocene period, but shells just like it can be found today, with living organisms inside them. Turritella is like a spiral staircase made from the top down with treads that gradually get bigger and bigger. The shape is a consequence of simple rules of growth.

As well as adding a new chamber on the end of the old one, and changing its size in a regular manner, the creature builds the new chamber at an angle to the plane of the previous one.

FIBONACCI'S FLOWERS

Some very characteristic patterns prevail throughout much of the plant kingdom. Many flowers consist of many more or less identical petals arranged in a symmetric ring around a central seedhead. Exceptions, the most striking being orchids, are often left-right symmetric. Symmetry is a key mathematical concept in the study of patterns, and it will play a key role in our quest to discover the shape of a snowflake. Plant life also exhibits some of nature's most striking numerical patterns. They are rules of thumb rather than ultimate truths, but it turns out that they embody some surprising consequences of the way plants grow.

Leonardo of Pisa, nicknamed Fibonacci, was born in 1170, the son of a customs officer. As a young man he joined his father in the customs house, where he was exposed to the new system for writing numbers invented by the Arabs and Hindus—the forerunner of today's decimal system with its symbols 0, 1, 2, 3, 4, 5, 6, 7, 8, 9. Impressed, he wrote *Liber Abbacci*, an arithmetic text. (In those days "abacus" with one b, referred to the apparatus with beads still used today for arithmetical calculations, but "abbacus" with two b's referred to the process of calculation.) This book, which dates from 1202, was the first to introduce Hindu-Arabic numerals to Europe.

About half of it is given over to calculations related to foreign currency exchange, but in among the more prosaic examples is a problem that has spawned a huge amount of mathematics. Ostensibly it is about rabbits. Start with a pair of immature rabbits (one male, one female). After one season they mature, nature takes its course and this one pair gives rise to a new pair of immature rabbits. In successive seasons all of the mature pairs give rise to one new immature pair, and all pairs that had been immature for the previous season mature and give rise to one new immature pair. All rabbits are assumed to be immortal. How does the population of rabbits grow? Just a little thought reveals a pattern. In successive years the number of pairs is: 1, 1, 2, 3, 5, 8, 13, 21, 34, 55, 89, 144…

and so on, where after the first two, each number is obtained by adding together the previous two numbers in the sequence.

Leonardo's rabbit population figures came to be known as Fibonacci numbers. Of course, rabbits don't reproduce according to Leonardo's rules and they don't live forever. Nonetheless, more sophisticated versions of Leonardo's scheme are used today in the study of animal populations. The Fibonacci numbers, though, have penetrated deep into the mathematical psyche as an apparently endless source of inspiration and wonder.

Plant numerology is riddled with Fibonacci numbers. Lilies have 3 petals, buttercups have 5, delphiniums often have 8, corn marigolds have 13, asters have 21, and daisies and sunflowers typically have 34, 55, or 89. Some bigger sunflowers have 144 petals. Numbers that are not Fibonacci numbers can occur in flowers, but these are less common, and most exceptions are either twice Fibonacci numbers (usually a consequence of plant-breeding techniques that double the number of petals) or belong to the similar anomalous sequence 1, 3, 4, 7, 11, 18, 29… with the same rule of formation but a different

RIGHT There are secret regularities to plants, and once you know what to look for, you can often find these regularities in unexpected places. The things to look for are spiral patterns, and special numbers that are closely associated with those spirals, called Fibonacci numbers. These patterns can be found in many plants, like fir cones.

starting point. If you get one petal less than a Fibonacci number, you can bet one fell off.

Fibonacci numbers also occur in the spiral structures that many plants use to arrange their seeds. Fir cones are a good example. The scales on fir cones are typically arranged in two families of intertwined spirals, and each family contains a Fibonacci number of spirals. Thus the Norway spruce cone has 5 spirals of scales in one direction and 3 in the other, while the larch has 8 and 5. The seedhead of a sunflower displays these spirals in a gloriously regular pattern. In small sunflowers there are 34 spirals in one family and 55 in the other; in larger ones these numbers may be 55 and 89, or even 89 and 144.

LEFT Fibonacci numbers are also found in flowers. Because biological growth is very flexible, there are some exceptions to these patterns, but Fibonacci numerology and spiral geometry are surprisingly common. They suggest that plant growth obeys simple but subtle mathematical rules, which lie somewhere in the interface between dynamics, geometry, and arithmetic.

8 13 21 34 55 89 144

DEEP SPACE

The puzzle of patterns extends beyond the bounds of planet Earth. Modern science came into existence because people noticed patterns in the heavens, and the first significant flowering of our mathematical understanding of nature occurred when we started to comprehend the cosmos. Ancient humans observed the lights in the night sky, wondered what they were, and then invented cosmologies to explain them. The Babylonians saw the heavens as a solid vault resting on the ocean—the Sun lives above the vault with the other gods and emerges every day through a doorway. The Egyptians saw the heavens as a flat ceiling, and the stars as lamps. The fixed stars hung on cords, while the "wandering stars"—the planets—were carried around by gods. These explanations are quaint to us, but they are attempts to come to grips with very real, and important, patterns of movement.

Some cosmological patterns are more obvious than the motion of the planets. The Earth is a (slightly flattened) sphere. So are many other bodies in the solar system—the Sun, the Moon,

Mars, the rest of the planets. The biggest asteroids—Ceres, Pallas, Vesta, Juno—are also round. Many of the smaller asteroids are more like giant potatoes, though—Castalia looks like a dog's bone. Saturn is surrounded by rocks and ice arranged in another mathematical shape—rings. Originally Saturn's rings were thought to form an annulus, a flat circular disk with a circular hole in the middle, like a washer. Then the disk was found to have gaps, also circular in shape. When the Voyager spacecraft reached Saturn they sent back images revealing that the fine texture of the rings is so intricate that it defies complete description; it is almost like the grooves on an old vinyl gramophone record, only far more densely packed. Some rings are braided, some have gaps, and some aren't even circular. Nonetheless, the geometry of Saturn's rings is dominated by circular structures. Saturn is not unique in this respect. Jupiter, Uranus, and Neptune also possess rings, but they are much fainter, contain much less material than Saturn's, and are often incomplete arcs.

The Sun is a star powered by nuclear reactions that make a hydrogen bomb seem like a whisper.

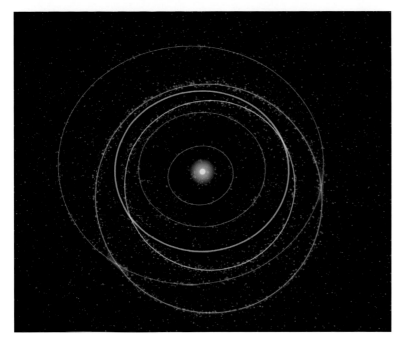

FAR LEFT Saturn's rings are not just a flat disk: they have gaps, and a very delicate structure. With a few rare exceptions, the rings have circular symmetry—for example the gaps have the same width all the way around.

LEFT Because gravity has the same effect in all directions, it arranges ice and rock into circles. Just like we see in Saturn's rings, approximate circular geometry also characterizes our solar system, for very similar reasons.

The light output of some stars is variable, often following a repetitive cycle of brightening and dimming. Our Sun's output isn't like that, though it's not constant either. The Sun undergoes another kind of variation—it vibrates. It "rings" like a bell when struck by starquakes, just as our own planet reacts to earthquakes, however, we need very sensitive equipment to observe the vibrations. When we do measure them, we see geometric patterns. The Sun is also speckled with sunspots, which are gigantic magnetic whirlpools that come and go in a cycle lasting roughly 11 years. Kepler also discovered more subtle patterns in the cosmos, resolving the old questions about wandering stars. He inherited many years of detailed and accurate observations of the movements of planets, made by the Danish astronomer Tycho Brahe in the 16th century. Being a pattern seeker, Kepler wasn't satisfied with a huge table of numbers. Instead, he looked for hidden hitherto unnoticed regularities. In his *New Astronomy* of 1609 he extracted two valid patterns of planetary motion from Brahe's data. That the orbits of planets are ellipses, with the Sun at one focus, and the line joining the planet to the Sun sweeps out equal areas in equal periods of time. In 1619 in *The Harmony of the World* he stated a third: the square of the orbital period (the time it takes a planet to complete one orbit) is proportional to the cube of the planet's distance from the Sun. The wanderings of the planets against the backdrop of "fixed" stars, an aspect of the heavens that had previously looked capricious, turned out to have hidden patterns. We have since found many other patterns in the heavens, some temporal and some spatial. Among the most amazing are gigantic spirals—we call them galaxies. Galaxies come in other shapes, but the most common shape has two spiral arms. A spiral galaxy is a slowly turning pinwheel made up of hundreds of billions of stars. We don't fully understand galaxies; in fact, we don't fully understand rather a lot of what we see through our telescopes. This shouldn't come as a surprise—ignorance, after all, is the human condition. The triumph of science is not yet Ultimate Knowledge, it is that every day we understand a little more about nature and the universe than we did the day before.

3

WHAT IS A PATTERN?

Kepler's three laws of planetary motion (*see pp. 26–7*), distilled from the mass of surrounding nonsense, led Isaac Newton to a dramatic insight—a single unified law of gravitation. Kepler's laws are equivalent to a mathematical rule for the gravitational force between any two bodies, anywhere in the universe. It's a simple rule. At twice the distance, the force of gravity is one-quarter as great, at three times the distance it is one-ninth as great, and so on. Twice the mass produces twice the force and three times the mass produces three times the force.

BELOW AND RIGHT Up There and Down Here. The movements of the stars and planets (below) can be reduced to simple, universal mathematical rules: the laws of motion and gravity. The movements and motives of human beings (right) are not so easily reduced to mathematical form. Yet throughout history new discoveries in either domain have often led to similar discoveries in the other. Up There and Down Here are a lot closer to each other than we imagine.

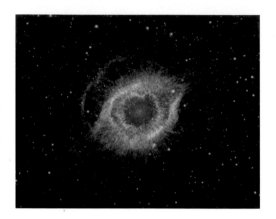

Newton's law of gravity was one of a select group of big discoveries that convinced 18th-century savants that we inhabit a clockwork universe. Once the universe has been set going, its entire future follows inevitably from immutable mathematical rules. This philosophy is known as determinism: it holds that in principle all events are predetermined, even if in practice we don't know what that predetermined outcome is going to be.

Human experience down on the ground doesn't seem to have the same kind of regularity—hidden or not—that goes on up in the sky. Human affairs seldom follow regular patterns; people please themselves, with unpredictable consequences. Indeed, a central aim of what we call "the law" is to regulate human affairs. We act as if we live in a lawless world, where we create regularity by imposing the rule of law. This is the exact opposite of the Newtonian philosophy.

Are we deluding ourselves, then, when we claim to detect mathematics at work behind the scenes of our universe? Or is there something special about the human level, something that

subverts or conceals the underlying lawfulness of events? Is all of nature founded in mathematical rules, or are we merely selecting those aspects of the natural world that happen to resemble human mathematics and then assuming them to be fundamental when in fact they are unrepresentative?

For that matter, everything that we know about the universe is conveyed to our minds through the medium of our senses. Our brains receive signals from the eyes and process them into "Tiger!" or "Wasp!", prompting appropriate evasive action; this indicates that our senses have in part evolved for the purpose of detecting patterns. So good are they at this task that we think we can see patterns where none exist, such as the Great Bear and the Swan in the night sky, which are really random alignments of physically unassociated stars. Maybe nature's alleged mathematical basis is a figment of human imagination.

I think there is a degree of truth in that last statement, but it isn't the whole truth. I am convinced that the specific mathematical structures that we claim to observe in nature

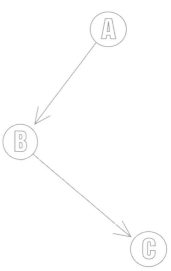

owe a lot to our own peculiarities and limitations. Intelligent plasma-vortex aliens living in the photospheres of stars would probably have no concept of numbers or triangles, but I bet they'd have far more sophisticated concepts of fluid flow than we do. A story that is simple enough for an overdeveloped ape brain to understand is unlikely to be the whole story. Nonetheless, the stories of science are pretty good stories—good enough for the apes to build heavier-than-air machines that fly across the oceans and forests, good enough to land overdeveloped apes on the Moon, and good enough to work out the contents of the apes' genetic instruction book.

For me, the possibility that reality might be a figment of human imagination pales into insignificance beside the fact that our minds are figments of reality. Mind is a system of interacting processes going on within a brain, carried out by ordinary matter obeying the same laws as all other ordinary matter. It isn't a problem if the rules upon which these processes run are much simpler than the processes themselves—the complexity of mind arises from the complex organization of these interactions. So it seems to me that the mathematical inclinations of the human mind are evolutionary responses to genuine patterns in the surrounding universe. Mathematicians won't work in a universe that bears no relation at all to mathematics.

BELOW Kepler's laws of planetary motion tell us that (1) a planet's orbit is an ellipse with the Sun at one focus (*New Astronomy*, 1609); (2) the line from the Sun to a planet sweeps out equal areas in equal times (*New Astronomy*, 1609); and (3) the square of the period of the orbit is proportional to the cube of the mean distance from the Sun. (*The Harmony of the World*, 1619).

1

2

3

ORDER & DISORDER

The development of science in the second half of the 20th century has made it clear that "pattern" is a subtle concept. As we develop new ways to extract hidden structure from our observations, this concept is constantly changing. Computer enhancement of blurred images, for instance, is performed using mathematical patterns that the unaided human eye cannot detect—that's why we see blurs when we look at them. None of this implies that "pattern" has ceased to be a meaningful concept. What it implies is that the section of nature's pattern book that is accessible to humans keeps getting bigger. We just have to learn new ways to look.

Another related term is "order," and along with that comes its opposite, "disorder." Not so long ago, the interpretations of both terms were considered so evident that it wasn't worth trying to make them precise. The distinction between order and disorder was as obvious as that between night and day. "Disordered" meant "random," and "random" meant there were no patterns. So it was pointless to look for patterns in disordered data. Now we find that there is a vast, largely unexplored, twilight zone between the midnight of disorder and the high noon of order. Between regular patterns and random jumbles there stretches a huge spectrum of more or less ordered, more or less random behavior.

Before we can appreciate just how revolutionary this new viewpoint is, though, we need to come to grips with simpler and more obvious kinds of patterns. These will open up an entry route into the twilight zone.

The simplest patterns that we can perceive are numerical ones. The ancient Greek followers of the Pythagorean school believed that the universe was driven by numbers—whole numbers, 1, 2, 3, 4, 5 and so on. They endowed each number with mystic attributes—

2 was male, 3 female, and their sum 5 signified marriage, for instance. The Pythagoreans supported their philosophy with a lot of irrational mysticism and numerology, but also with some astute observations of the natural world—such as numerical patterns that occur in musical harmony, to which we shall return (*see pp. 120–121*). Not only that—they noticed key patterns in the structure of numbers themselves.

Some of their simplest and most striking patterns arise by counting geometrical arrangements of objects. The square numbers 1, 4, 9, 16, 25—a term we still use—count objects arranged in squares:

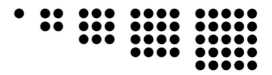

The triangular numbers 1, 3, 6, 10, 15, 21 similarly count objects arranged in triangles, like snooker or pool balls at the start of a frame:

A typical Pythagorean discovery is that if we add consecutive triangular numbers, we get squares: 1+3 = 4, 3+6 = 9, 6+10 = 16, 10+15 = 25. This pattern can be explained by some clever geometry. Split a square arrangement along a diagonal and we get:

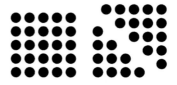

The Pythagoreans took the first steps toward the concept of "proof" in mathematics. They also recognized that there are connections between arithmetic and geometry, even though the raw materials of these two branches of mathematics seem so very different. One of my favorite "patterns" of this kind—it is almost a coincidence, except that no property of numbers can be truly coincidental—arises if you pile spherical cannonballs in a square pyramid. The top layer contains one cannonball, the next layer contains four, and so on—successive squares. The totals go 1, 5, 14, 30…and by the 24th layer, the total number of cannonballs reaches 4,900, a square number. You could lay out all 4,900 of them in a 70 by 70 square if you wanted to. It turns out that the 24th layer of the pyramid is the only one (other than the very first) for which the total number of cannonballs is square. This numerical fact was noticed long ago, but the first proof was obtained in the 1930s. I don't think it's terribly significant…but it's fun.

ABOVE AND BELOW Sometimes the universe exibits a clear, simple pattern: we call this order. At other times, everything seems jumbled and irregular: we call this state disorder. But order and disorder are human ways of classifying the world, linked to features of our own perceptions and minds. When nature makes a spiral galaxy (above), or an irregular one (below), the laws involved are identical. Only the circumstances have changed.

SYMMETRY

So far, my aim has been to display the diversity of nature's patterns, but now we can start to unify them. The most important ingredient in the current mathematics of patterns runs like a golden thread through everything we have seen so far. That thread is symmetry.

In everyday language, the word "symmetry" is often used rather loosely, to indicate elegance of proportion. In mathematics, the word has a precise meaning. A shape is symmetric if it looks exactly the same after being transformed in some way—reflected, rotated, slid, expanded, contracted. Each such transformation is called a symmetry of that object.

The simplest mathematical example is bilateral symmetry, in which the left and right sides of an object are identical save for a reflection. The human form has approximate bilateral symmetry externally, though this is less exact than we often assume. Faces appear symmetric at a casual glance, but a closer look usually reveals that one eyebrow has a slightly different shape than the other, or one corner of the mouth turns down slightly compared to the other. Artificial faces made by taking half of a person's face and joining it to its mirror image are usually instantly distinguishable from the individual's actual face—and the two halves give different results.

Bilateral, or left-right, symmetry (again, on the surface if not within) is the norm in the animal kingdom. Butterflies are symmetric, each wing bearing a mirror image of the pattern on the other. Dogs, cats, cows, goats, sheep, horses, elephants, camels, hedgehogs, pigeons, swans, swallows, lizards, frogs, beetles, moths, spiders, lobsters, coelacanths, manta rays, sharks—all are more or less left-right symmetric. The fossil record shows that bilateral symmetry has been around for a long time, certainly since the Ediacaran worm Spriggina, which lived some 560 to 580 million years ago.

There is more to symmetry than just left-right reflection, however. The commonest starfish have five roughly identical arms, equally spaced to form a star. The most obvious symmetry here is rotational. That same starfish can be rotated into five different positions, and in each of which it looks exactly the same as it did to start with. Flowers also often have rotational symmetry. And so does a snowflake.

Starfish also possess bilateral symmetry. If you imagine a line running along the middle of one arm and separating the other arms into two adjacent pairs, then the shape to the left of this line is the same as that to the right. In fact, there are five such lines—one for each arm. Again, the same generally holds for flowers and snowflakes.

Why is symmetry—especially bilateral— such a prominent feature of so many living organisms? Is it a consequence of how creatures grow? The developing embryo—of, say, a frog or a newt or a human—undergoes several changes of symmetry, but from a very early stage there is a dominant bilateral symmetry, which thereafter

is maintained except in such details as the positioning of various internal organs. Perhaps bilateral symmetry is a consequence of the growth process. After all, if the left half of an animal grows according to certain rules, and the right half obeys the same rules, then both halves ought to look the same. Conversely, developing organisms may have to work quite hard to keep themselves symmetric as they grow because slight departures from left-right symmetry can easily be amplified into major differences as the creature develops (and sometimes are). The starfish also changes symmetry as it develops. It begins as a bilaterally symmetric form, and the part that eventually becomes fivefold symmetric first appears on one side only; then the rest of the organism is used up or discarded and only the fivefold symmetric part continues to develop. Symmetry in living creatures retains much of its mystery.

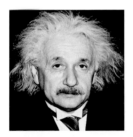  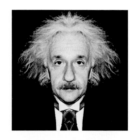

LEFT AND BELOW LEFT
A starfish has five reflectional symmetries (below left), while Albert Einstein only has one (left). The symmetries are imperfect, so that fitting together two copies of Einstein's left or right half give very different results.

SELF-SIMILARITY

Symmetry is not the only important general concept to have been abstracted from nature's patterns. There are also other, less obvious, kinds of regularity.

A friend of mine has a photograph, of which he is very proud, taken on vacation in the Norwegian fjords. It shows him standing up in a small boat, nonchalantly leaning one elbow on a ledge of rock—as if the boat is moored at the water's edge. Actually, the rock is 1,650ft (500m) high, a steep cliff forming part of the fjord's rocky face. My friend is in the foreground, while the rocky ledge is half a mile (kilometer) or more away in the background. Nonetheless, the illusion is convincing. It is effective because of a hidden pattern in rocks, one that mathematicians have only recently discovered and analyzed.

The pattern is deceptively simple—small rocks viewed close up look just like big rocks seen from a distance. You can't detect this kind of pattern just by looking at one rock. It is a collective pattern common to most rocks when they are viewed on many scales of magnification. It is known as self-similarity. In fact, for rocks, what the human eye picks up is statistical self-similarity. Small bits of rock have the same kind of random texture as big bits. Clouds, mountains, coastlines, and craters on the Moon are also statistically self-similar. This is important because it tells us something about the processes that create those forms. Those processes must work in the same manner on many different spatial scales.

Some objects have a more regular kind of self-similarity—small parts of them are miniature versions of the whole thing. In nature, this kind of self-similarity is never perfect. The small pieces closely resemble the whole object, but differ from it in detail. Perhaps the best example is a fern. A fern consists of a series of fronds, ranged either side of a central stalk. The fronds are largest at or near the base of the stalk and they taper off as they approach the tip, giving the fern its characteristic soft triangular shape. The same description applies to each frond. It consists of a series of sub-fronds, ranged either side of a central stalk; the sub-fronds are

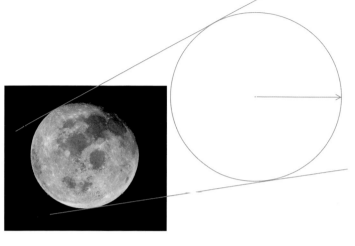

largest at or near the base of the stalk and they taper off as they approach the tip. On many ferns, the sub-fronds also have the same kind of structure as the fronds and the fern itself, though the detail often starts to become rather sketchy, as if nature couldn't be bothered to make such an intricate structure yet again. On other ferns, however, it is possible to find a fourth generation of frondlike structures.

Mathematicians capture the patterns of nature by constructing ideal forms—clean and tidy versions of nature's less regular structures. In mathematical astronomy, for instance, the Moon, which has craters, mountains, and is flattened at the poles, is idealized as a mathematical sphere. A mathematical sphere is perfectly smooth. Every point on its unblemished surface lies at exactly the same distance from its center. If you could measure that distance to a trillion decimal places, every single digit would be the same for every point on the surface of the sphere. Nothing in nature has that degree of accuracy, but it makes mathematicians' lives far simpler to pretend that all of those distances are equal.

In this spirit, mathematicians idealize nature's approximate self-similarity into exact self-similarity. A mathematical shape is self-similar if it can be assembled from several smaller-scale copies of itself, perfect in every detail. So a mathematician's fern goes on forever, and every tiny sub-sub-sub-frond is precisely the same as the whole fern—just a lot smaller. A similar trick is used to idealize statistical self-similarity—the smaller copies should have exactly the same statistical distribution of features as the whole.

In particular, the mathematician's fern has detailed structure on scales far smaller than an atom, or even the smallest meaningful length scale in the physical universe, the Planck length—roughly ten trillion-trillion-trillionths of a meter. No matter. The mathematician's idealization merely tidies up an approximate regularity of nature, taking it to an extreme—but a meaningful and a useful extreme. The idealization captures the important properties of the reality—and it's a lot easier to think about the ideal case than the imperfect reality.

DYNAMICS

Most of the patterns I've mentioned so far are fixed, static. Or, at least, they seem that way if you look at them for a few minutes, even though they may change over longer periods of time. Plants and animals grow, sand dunes advance across the desert. At White Sands in New Mexico the gypsum sand extends its boundaries by several yards every year. On a windy day you can see the sand moving. Clouds of windblown dust swirl about and occasional tiny avalanches slip down the steep faces in the lee of the dunes.

Other patterns are more dynamic and can be detected only by observing how they change over time. Examples are the increasing number of people alive on the face of the Earth, the climate changes associated with global warming, or the puff of mud created in a pond by the sudden flick of a fish's tail. Any system that changes with the passage of time is called a dynamical system. The changes that occur are the system's dynamics.

The ancients discovered dynamical patterns in the night sky. Every night the stars of the northern hemisphere seem to rotate around the Pole Star through an angle of 15 degrees every hour. (In the southern hemisphere they also rotate, but there isn't a bright star sitting close to the axis of rotation.) Every month the Moon changes its apparent shape, waxing from new to full and then waning back to a thin crescent, before starting the cycle again. Of course it's not the actual shape that is changing, but the apparent shape of the Moon's sunlit side. By observing the phases of the Moon, the ancient Babylonians and Greeks worked out that our sister world is a sphere and that it shines by reflected sunlight.

There are dynamic patterns down on the ground, too, but on the whole the ancients were much less aware of them. Patterns of animal movement—the pace of a camel, the trot of a horse, the amble of an elephant, the gallop of a charging rhinoceros. Patterns of weather, such as anticyclones, where air and clouds swirl in gigantic spirals. In the tropics, some spirals in the atmosphere feed on heat and moisture and grow into roaring hurricanes, sweeping aside everything in their path with winds up to 125mph (200km/h). To the ancients, storms were capricious acts of willful intent—the whims of the gods. Today we know that the gods of weather are bound by mathematical rules, but only some consequences of these rules are predictable by humans, even using powerful computers, so it is still impossible to predict the weather more than a few days ahead.

BELOW The phases of the Moon and a galloping horse both repeat the same cycle of changes over and over again. Here the patterns appear in time, not space.

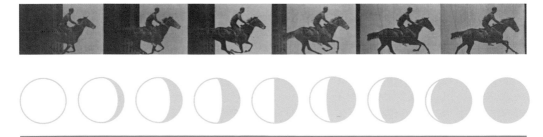

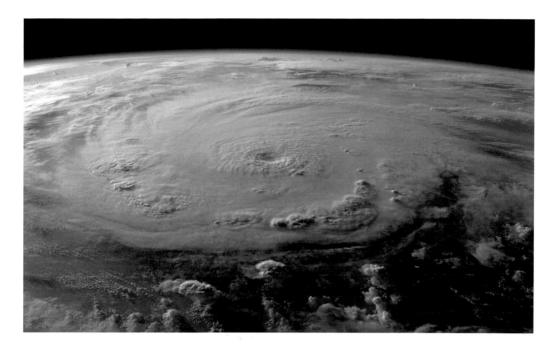

All this tells us that the true patterns of dynamical systems lie in the rules. The patterns we observe in the effects of those rules are cryptic clues to the rules themselves. Sunshine, cloud, rain, hail, snow—all these derive from the same small set of rules, which we express as mathematical equations. The rules of nature are elegant, and so are some of their consequences, such as raindrops in a puddle, aspens shivering in a sudden breeze, salmon-pink stratus clouds at sunset, cresting snowdrifts—and snowflakes.

Other consequences of these rules, though, have no immediately obvious elements of pattern or structure—a shower of rain, a wind-battered field of corn, a howling blizzard, a stormy sea. Yet the underlying rules are just as pretty as those for a snowdrift or a raindrop, for they are the same rules.

It is this paradox that excited Newton and his successors. Nature operates on many levels. What seems incomprehensible on one level may

ABOVE Some of the most interesting of nature's patterns are patterns of movement or change. Seen from close up, a hurricane is violent and erratic. When viewed on a global scale, it reveals itself as an elegant spiral of air and moisture, spinning in stately fashion over the surface of the ocean.

become obvious—or at least understandable—on another. Even patterns that can be given an elegant description on one level, the level of things, may still be best explained in terms of rules that operate on a deeper level, the level of processes. This is how science has developed, starting from human perceptions of the world, but seeking deeper explanations through "laws of nature"—mathematical rules that capture certain regularities in our universe.

If we are to understand the shapes of snowflakes, we must first understand the processes that give rise to them—and the laws that mold these processes.

PART TWO

THE MATHEMATICAL WORLD

4
—

ONE DIMENSION

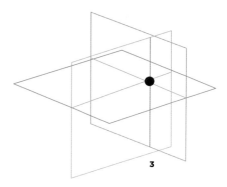

3

In order to provide our understanding with some firm foundations, we're going to construct a catalog of patterns. If this catalog is going to be useful, there has to be some organizing principle. Mail-order catalogs organize their contents into broad categories—jewelry, kitchenware, CDs, toys. Ours will do the same, but with different categories, including dimensionality, symmetry, and continuity.

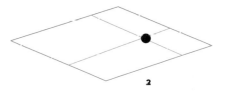

2

The dimension of a mathematical space, roughly speaking, is how many numbers it takes to specify whereabouts you are. The surface of the Earth is two-dimensional. You know where you are provided you know your latitude and your longitude. If you leave the Earth's surface, then a third number, your height, is required, so space is three-dimensional. Most of Euclid's geometry takes place in the plane, which is two-dimensional. Any point can be specified by two numbers—how far it lies along an east-west axis, and how far it lies along a north-south one. A line is one-dimensional—just use an east-west axis. Even simpler is a single point, which on its own is zero-dimensional—you don't need any numbers to say where you are, because you can only be in one place anyway. A zero-dimensional point is a bit too simple to be interesting, but one-dimensional spaces have a lot more going for them than you might expect.

1

For instance, think about a line of footprints in the snow. If the person making them is walking in a regular rhythm—say on level ground, no snowdrifts, the simplest "default" movement when the mind is otherwise engaged and the feet move almost of their own accord—then the footprints form a regular pattern of two evenly-spaced parallel tracks. One line of prints consists entirely of left feet, regularly spaced; the

0

ABOVE The dimension of a space is the number of independent directions that exist within it. The space that we inhabit has three dimensions, because it is possible for three lines to meet each other at right angles. A plane, a flat surface, has only two dimensions, and a line has only one. A space consisting of a single point contains no directions, and is therefore zero-dimensional. In conventional physical space, three dimensions is the limit, but mathematical spaces with four or more dimensions are just as easy to define and manipulate.

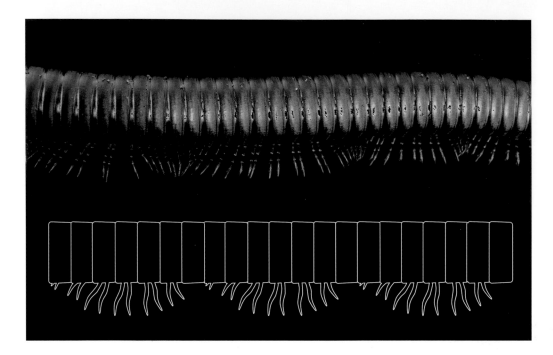

other line of prints consists entirely of right feet. The left-foot prints are not level with the right; instead, the left-foot prints fall into the gaps between the prints on the right. Similarly the right-foot prints fall into the gaps between the prints on the left.

Admittedly, the pattern in the snow is not quite one-dimensional because there are two lines of footprints and, anyway, feet have width as well as length. Nonetheless, the most important feature—the sequence in which left and right prints occur—is linear in character. All of the important action takes place along the direction of movement. So, slightly metaphorically, in the interests of our subject here I'll consider footprints to be one-dimensional patterns.

In fact, footprints fall into the general category of frieze patterns. These are formed by spacing copies of some shape (or shapes) along a line at regular intervals. To keep things interesting, the shape can be two-dimensional, but it has to stay on (or at least near) the line. In these terms, we can idealize footprints into the frieze pattern ⌐ ⌐ ⌐ ⌐ ⌐ ⌐ ⌐ ⌐

where U̱ represents "left foot" and ⌐ is "right foot." If the person walking places each foot in front of the other, the footprints would look just like this, but even if she doesn't, the idealization captures the symmetries of the line of footprints. A symmetry is a way to repeat a pattern, and here we find two different types of symmetry. The pattern of the frieze repeats if it is translated, or slid, two steps forward; this is the first type of symmetry. The second type of symmetry captures the left-right sequence in terms of a glide reflection: slide one step forward and then reflect left and right. The pattern does not look the same, however, if it is reflected left-right without the glide.

Frieze symmetries are combinations of translations, glide reflections, ordinary reflections (no glide), and rotations—swing the whole pattern around in the plane through 180 degrees. Two distinct reflections may occur— left-right and front-back. The most important distinction between friezes is their list of symmetries. Other distinctions, such as the actual designs used, have artistic significance, but do not affect the basic pattern.

LEFT The millipede advances smoothly, with the aid of a series of traveling waves that ripple along its legs. This creates a pattern that repeats in space as well as time.

RIGHT Unlike the millipede, the inchworm creates forward movement from a standing wave. With its front end fixed, it flexes its body into an inverted U and pulls the back end forward. Then the back end is anchored, the front end freed, and the U is flattened out again. The inchworm is now stretched out, as before, but has advanced by about half its body length.

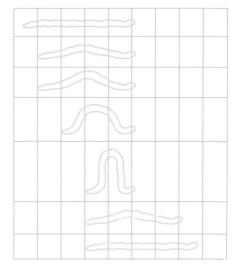

FRIEZE MOTION

Nature makes good use of frieze patterns, often where you wouldn't expect them. A centipede is a moving frieze. There are 2,800 species of centipede. The creatures have between 14 and 177 body segments, each bearing two legs. A mathematician's centipede would be infinitely long, with no head and no tail; just an endless series of identical segments, each with a pair of legs. This doesn't mean that mathematicians believe in unrealistic biology, just that they prefer to idealize problems in order to bring out particular features. When a centipede moves, its legs create traveling frieze patterns. The centipede's motion thereby combines dynamics and symmetry, making it an excellent prototype for thinking about patterns of movement.

Real centipedes wiggle slightly from side to side as they move, and one that's moving fast wiggles a lot. In the spirit of idealization, consider the wiggles to be part of the frieze "decor," important for the actual mechanics of centipede motion but a side issue when it comes to the underlying pattern. When viewed with the pattern-seeking eye of a Kepler, what features of the centipede's progress are most striking? Look first at the line of legs along the creature's left side. As it moves, each leg stretches forward and then flexes backward, just like the oars of a rowboat. And as with the oars, the effect is to lever the creature along. In a rowboat, all the oars move in synchrony, but nature chooses a different pattern for the centipede, one that avoids stressing all the legs at the same time, achieving smoother movement with less effort. The centipede's legs move one at a time, in sequence along its body. Waves of movement ripple along the creature's side. The waves start from the back and travel toward the front. At low speeds two or three waves fit into the centipede's length, and at higher speeds there are fewer waves—typically one and a half—that pursue each other along the creature's left side.

Meanwhile, the right side has not been idle. (If it had been, the centipede would have been traveling in circles.) The same pattern of waves travels along the right side, too. However, left and right do not move together. When a left leg is fully extended in the forward direction, the corresponding right leg is fully extended backward, and vice versa. Each segment behaves like a walking human, whose strides to left and right alternate. In the abstract, the centipedal frieze is the pattern | | ⌴ ⌐ ⌴ ⌐ ⌴ ⌐ ⌴ ⌐, but now ⌴ and ⌐ are waves.

Another myriapod, or many-legged creature, is the millipede, which generally has between 22 and 200 legs. There are roughly 10,000 species of millipede. A millipede moves much like a centipede, except that the legs on left and right are in synchrony. So the millipede's frieze is C C C C C C C C. One reason for the difference with that of the centipede is that millipedes are generally slower walkers.

Caterpillars have a lot of legs, too, although technically they aren't myriapods but juvenile forms of butterflies and moths. They have 13 segments, but not always bearing legs. Caterpillars are also mobile friezes, typically of the millipede's left-right symmetric kind. Sometimes their movement, while still fitting this pattern in the abstract, takes on a more bizarre appearance. On the other hand the inchworm has four legs at the front, then a long gap, then six legs at the rear. These six legs do multiple duty, standing in for the missing ones that ought to be in the gaps. As repeated waves of movement ripple through the inchworm's rear-engine drive, its tail marches up to its head and the body arches into an inverted U shape. When tail meets head, the rear legs secure themselves, the front ones let go, and the body suddenly straightens out with the head some distance forward from its previous position. Then it does the same thing again. And again.

WIGGLES AND WRIGGLES

Snakes, worms, eels, and lampreys also have bodies so elongated that the mathematician cannot resist the temptation to idealize them into a simple linear structure, although this time the idealization has no legs.

Sometimes this idealization is used primarily for motivational purposes, to dramatize an otherwise dull-sounding scenario. The notorious (and still unsolved) problem of Mother Worm's Blanket is a case in point. Baby Worm curls up in some random shape when she goes to sleep, a different shape every night. Frugal Mother Worm, always keen to keep family expenditure within budget, wants to make the smallest blanket—the one with minimal area—that is guaranteed to cover Baby Worm however she curls up. More prosaically—what is the minimum-area shape that can cover all curves of unit length? Nobody knows. But whatever the answer, the worm version of the problem sounds far more interesting.

Sometimes the mathematician's lineal worm has a more serious purpose. Although worms have no legs, their patterns of movement have a lot in common with millipedes. A worm's body has two series of muscles, sandwiched between its skin and its internal organs. One set, the circular muscles, consists of ring-shaped muscles that go around the body. The second set, the longitudinal muscles, runs lengthwise. When the worm moves, a ripple of contraction of the circular muscles flows along its body from front to back. When this ripple has got about halfway along the body, the longitudinal muscles contract. Like the legs of the millipede and the strokes of the oarsman, these waves of muscular activity propel the worm through the soil. Contraction of the longitudinal muscles thickens the worm and allows it to get a grip against the walls of the tunnel that it is boring through the ground. Contraction of the circular muscles elongates the worm's body, releasing it from the walls and pushing its free end forward. Again the waves are in synchrony on left and right—indeed, all the way around the body.

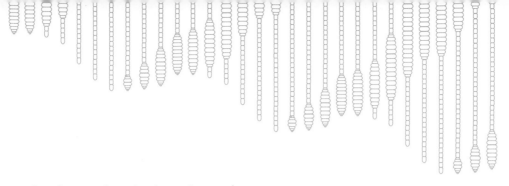

Snakes also move by activating major muscle groups in sequence, but here the pattern is closer to that of a centipede, with opposite sides out of synchrony. When a muscle group on the left contracts, its opposite number on the right relaxes; when a muscle group on the left relaxes, its opposite number on the right contracts. Like the centipede, snakes generally wiggle from side to side when they move, a form of motion described as serpentine. All sections of the body start to move at the same time and stop at the same time, and wherever the head goes, the rest of the body faithfully follows.

A second type of movement in snakes is concertina motion, used if the animal is confined within a narrow channel. It leaves part of its body free, while zigzagging the rest to and fro between the walls of the channel to brace it. By changing the points of contact in a manner analogous to a moving worm, the snake manages to make its way forward even though there isn't enough room for its normal serpentine flow.

Eels move in much the same way as snakes, but underwater rather than on land. The same goes for lampreys, which are technically not eels but look very similar. Many different creatures, then, are mobile friezes, and at first sight they move in a huge variety of ways. However, there is an inherent unity, a pattern that is common

ABOVE AND BELOW The earthworm burrows through the soil using a repetitive sequence of muscular contractions, which take place alternately around and along the length of its body. When the longitudinal muscles contract, the worm shrinks; when its circular muscles contract, the worm lengthens. These contractions also push the soil outward, leaving room for the creature to pursue its travels.

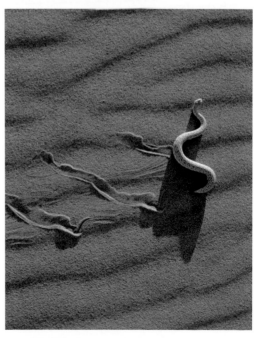

RIGHT Snakes use similar rhythmic contractions of their muscles to slither forward. The sidewinder employs an unusual variation: it flips its way across the desert, leaving a series of separate tracks in the sand.

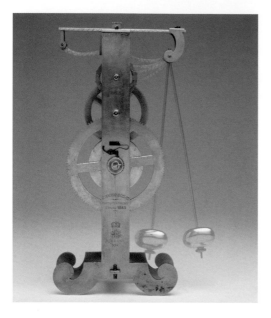

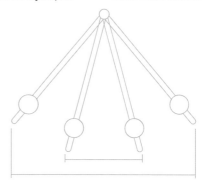

LEFT AND BELOW A pendulum clock ticks in a regular rhythm, and the more regular the rhythm, the better it keeps time. Sound is the result of rhythmic vibrations: bigger vibrations create louder sounds.

to them all, even the ones that have no legs. All these creatures move by setting up traveling waves of muscular contraction and relaxation. This tells me that if we're going to understand patterns of animal movement—our prototype for all dynamic patterns, including the making of that elusive snowflake—then it's not going to be enough just to look at legs. Quite simply worms don't have legs, yet they clearly make use of the same patterns as millipedes. Legs move because they are driven by muscles; worms keep the muscles but don't bother with the legs. But why do muscles move? How do muscles move? More crucially still—what is it that controls the timing of these muscle movements? To these questions, we shall return (see pp. 132–137).

CYCLES

Harold Macmillan, British prime minister from 1957 to 1963, when once asked what kept him awake at night, replied: "Events, dear boy, events."

An event is a change in the universe between one moment and the next. Without time, the universe would be a dull place. Nothing would happen. Time is one-dimensional—it can't run sideways. Movement is a sequence of changes of position, occurring in a way that is ordered by the forward flow of time. Dynamics is a temporally ordered sequence of changes of state; often the system's state is position, but it might be temperature, humidity, level of electrical activity, color, psychological mood, or even the price of fish.

Dynamical events can be erratic or they can be regular. Back in Renaissance Italy, a youthful Galileo was watching a swinging lamp suspended from a church ceiling and he noticed that the swings took the same time whether they were large or small. This gave him the idea for a pendulum clock. Ironically, his key observation is only an approximation to the truth. The time it takes a pendulum to swing does depend on the size of the swing, with big swings taking longer—but the error is surprisingly small, and in Galileo's day it was too small to matter. The wise clockmaker avoids this issue by making sure that the swings of the pendulum are all of the same size.

The regular swing of a pendulum is a simple instance of a periodic cycle—a dynamic that repeats the same behavior over and over again at equally spaced times. Patterns of animal movement provide further examples. Such cycles

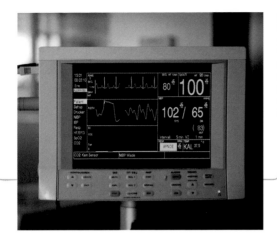

LEFT AND BELOW Our hearts beat rhythmically too, but not as regularly as a clock. The heart lacks the precision that we associate with machines. However, that is to our advantage because the heart has to respond flexibly: it must beat faster when we are running and more slowly when we are resting. Hospital monitors display the pattern of electrical activity of the heart, making it possible to diagnose heart disease by observing the timing of the heartbeat.

are evidence for the deterministic character of the laws of nature—Newton's idea that we live in a clockwork universe which, once set in motion, follows a predetermined course. A repetitive cycle occurs because a system that returns to its initial state must repeat whatever it did first time around—that's what deterministic means. If, however, the laws permit several behaviors in the same circumstances—which can happen if the laws involve chance—then there is no reason to expect the cycle to repeat.

In keeping with this observation, periodic cycles are an extremely important class of dynamical pattern, because it is easy to predict their behavior. It never ceases to amaze me, for instance, that the date of Christmas can be predicted years in advance! (So can the date of Easter, but not so easily now the date involves several competing cycles, together with some obscure ecclesiastical conventions.) The irresistible prediction for any periodic cycle is that next time around it will do exactly the same as it did last time.

We are surrounded by more or less periodic cycles—waves moving up a beach and breaking in the shallows, the daily movement of the Sun from one horizon to the other, not to mention the yearly cycle of the seasons. Birds migrate for the winter and return for the spring. Monarch butterflies make annual journeys of thousands of miles to their mating grounds. Wildebeest and elk do the same.

Periodic cycles form the basis of music, too. I don't mean the notes—they may sometimes be periodic, but not very often. No, I mean the vibrations that create the sound that produces the notes. When a violin string sounds middle C, it moves through a periodic cycle that lasts roughly 1/250 of a second. The same goes for the strings of a guitar, the air in a clarinet, oboe, or organ pipe, and the skin of a drum. Light, too, is a periodic series of vibrations, but these vibrations are much more rapid than sound, and what vibrates is electrical and magnetic fields.

Even our very lives depend on the successful replication of a periodic cycle. When we are resting, our hearts beat in a repetitive rhythm; when we become more active, the number of beats per minute increases to improve the flow of oxygen in our bodies. Our normal breathing is also cyclic in character, and when we walk, our legs move in a periodic cycle. (Indeed, when cycling our legs move in a periodic cycle and so do most components of the bike!)

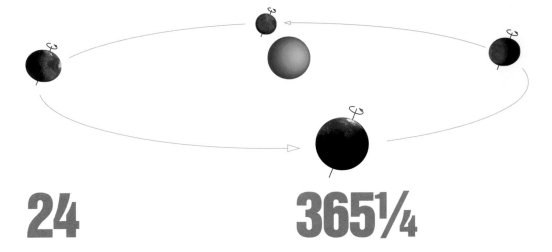

24

365¼

Sometimes it seems as if we, the entire planet and the universe itself is running on a huge, intricate system of interlocking cycles. Of course, it's not that simple.

CELESTIAL CYCLES

There are cycles in the sky, too, that also affect us on the ground. On the whole, the first astronomical cycles to be noticed didn't affect us much down here, while the ones that did affect us weren't thought to be related to the heavens. Eventually we got most of these things straightened out, but it took a good few thousand years.

The celestial cycle that has the greatest effect on life on Earth is the day-night cycle, which takes 24 hours to repeat. We now know that this cycle results from the rotation of the Earth, but to early humans it was obvious that the Earth was standing still (for heaven's sake, you could see that it was; anyway, if it was moving we'd all fall off). Later it dawned on people that everything would look like that no matter what the Earth was doing because the observers were being carried along with what they were

observing. Nonetheless, the belief that the Earth was the center of the universe and that everything else, including the Sun, went around it took a lot of effort to dislodge.

There are innumerable other celestial cycles. The cycle of the seasons repeats roughly every 365¼ days, taking the rising Sun through all the 12 constellations of the zodiac in turn. We probably owe these constellations, give or take the odd star, to the Babylonians, who were avid observers of the night sky. They made tables of the motion of Jupiter, for instance, and recognized cycles in that. Indeed, the ancients made such accurate observations that by the time of the Greeks they were aware of the precession of the equinoxes—a slow drift in the date on which day and night have the same length—a cycle that takes nearly 26,000 years to complete.

Nowadays we are aware of just how insignificant our planet is in relation to the true vastness of the universe. In the great universal scheme of things this is a not very distinguished world that circles a not very distinguished star in a not very distinguished galaxy, but it is our own.

LEFT Some of the earliest of nature's patterns to be recognized are of astronomical origin—the yearly cycle of the seasons (far left), the Moon's monthly cycle of phases, and the wandering movements of the planets among the "fixed" stars. Ancient civilizations were quick to notice these patterns. More modern astronomical patterns include galaxies (left) and black holes.

26,000

There are cycles even on the galactic scale—our galaxy is a giant whirlpool of stars and our Sun revolves once around the center of the galaxy every 240 million years.

As always, the mathematical idealization captures some aspects of reality but ignores others. The day-night cycle may last 24 hours, but the pattern of light and darkness varies with the seasons. Near the poles in so-called lands of the midnight sun the 24-hour cycle is not distinguishable by periods of day and night. The Sun stays above the horizon for almost quarter of the year, below it for another quarter, and light and darkness alternate during the periods between. These complications arise from the tilt of the Earth's axis; the 24-hour period is really the period in which the Earth rotates about its axis, and the cycles of day and night are consequences of that rotation—plus small effects from its motion around the Sun.

Periodic cycles in the heavens can have a profound effect on life on Earth. Inside many organisms is a biochemical clock, which stays roughly in synchrony with the day-night cycle and indeed resets itself in response to the pattern of light and darkness. Life on Earth dances to the solar rhythm. And the motions of the Sun and Moon relative to the Earth are the main factors that determine the timing of the tides and, to a lesser extent, how high or low they are. (The height also depends crucially on local geography and the weather.) What causes the tides? As the Earth rotates, lunar gravity causes the oceans to rise slightly on the side nearest the Moon, though not exactly beneath it. Contrary to intuition, they also rise on the side farthest from the Moon. The reason, roughly speaking, is that tides are mostly the effect of sideways movement of water, not vertical movement. Many living creatures exist only because of this flow of the tides, which gives them a new ecological niche—the intertidal zone—cycling from wet to dry and back again roughly every 12 hours.

Our earthly world has evolved within a far greater piece of machinery, and it responds to that machinery whether or not human beings are smart enough to notice. "Up there" and "down here" are not as far apart as we like to think.

5

MIRROR SYMMETRY

The most familiar example of symmetry is also one of the most puzzling—the mirror. The world that you see reflected in a mirror is still entirely convincing, but totally unreal. Mirrors hint at the existence of a second, "virtual" world that mimics our own, but with tantalizing differences. We can't get into that world, yet it interacts in some weak fashion with ours, influencing how we adjust a tie or put on makeup, for example.

We are right to be puzzled by mirrors not only because they are optically confusing but because this same mirror symmetry lies at the very heart of fundamental physics, and also hints at the existence of a virtual world. A world much like our own, even interacting with it, albeit only weakly and subtly but a world that is emphatically different.

Quite simply a mirror is a surface that reflects light. When you look into a mirror and see an image of your own face, the light has traveled from your face to the mirror and bounced back to enter your eyes. We are so used to this happening that we seldom ask why the image is so crisp and undistorted. The answer depends on a slightly different meaning of the word "reflect" from the one we usually use. In mathematics, you reflect an object in a plane by (conceptually) moving every point in it to the corresponding position on the far side of the plane. In other words, a mathematician's reflection actually does what a mirror's reflection seems to do. It places the object in the position that it would occupy if the mirror were a window and you were looking through it into the mirror world. A mathematician's reflection therefore flips the real and virtual worlds into each other.

The geometry of light rays is the same in both the real and virtual worlds (thanks to the deep mirror symmetry of physics), and that's what makes the reflected image so convincing.

But also baffling. Place a right shoe in front of the mirror, and it creates an image that appears to be a left shoe. Raise your right hand, and your mirror image raises its left. Apparently mirrors swap left and right. Why do they not swap top and bottom? In a mirror, your head remains at the top and your feet stay on the ground. Would this change if the mirror lay on its side?

We are baffled by mirrors because we are bilaterally symmetric. There are two ways to match the mirror image of a person with a real person. The human visual sense evolved in a world where objects can be moved, but not reflected (in the mathematician's sense). So subconsciously we rotate the image in a mirror, compare it with our real selves, and think that left and right hands have swapped. Actually, all parts of the reality and the image are in the same relative locations—head at the top, feet at the bottom, left hand on the left (your left!), right hand on your right.

What has changed, then? Not left and right, but front and back. You face north; your image faces south. A mathematical reflection squashes an object flat, passes it through itself, and opens it up the other way around.

It's much the same with shoes. We interpret the reflection of a right shoe as if it had been rotated, in which case it must be a left shoe. But it has been reflected, not rotated. What possesses "handedness" is not objects, but space; and the virtual space as seen in a mirror has the opposite handedness to the real space of which it is a reflection. Within the mirror world, a left shoe is actually a right shoe, that left hand a right hand.

If so, a second reflection should restore handedness. If you set two mirrors at right angles, and look into the result along the

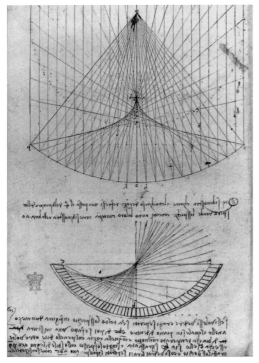

diagonal, you see your face, bisected by the line along which the mirrors join. Raise your right hand. Your image raises its right hand—that is, what looks to you like its right hand. This is very disconcerting, because you expect to see it raising what looks like its left hand. But it doesn't. The image you see has been reflected in both mirrors, and its handedness is therefore the same as that of the real world.

BILATERAL SYMMETRY

The animal kingdom is dominated by one kind of symmetry, the simplest kind: bilateral symmetry. The external form of bilaterally symmetric creatures is essentially unchanged when they are reflected in a mirror. Many plants also have this mirror symmetry, for example in their leaves. Orchids are a glorious, exuberant example of flowers that nearly all have striking bilateral symmetry but lack the rotational symmetry of daisies, dahlias, sunflowers, and the like. Any explanation of bilateral symmetry in

living creatures must take care of the exceptions as well as the general rule. A complete explanation has to take modern discoveries in genetics into account, as well as the dynamics of biological development. It must also be consistent with the fact that, to some extent, bilateral symmetry is not all that it seems. The internal organs of the human body are not as symmetric as the exterior. The heart usually lies on the left, and the convolutions of the intestines have a specific handedness. The left lung has two lobes, the right has three.

There are a good number of mechanical reasons why the intestines must be asymmetric. The intestine is a tube, which starts and finishes in a roughly central position in the body. If such a tube were arranged with left-right symmetry, it would have to wind around in the body's central plane without deviating to the left or right. But in order to digest food, it has to be long enough—too long to fit in such a position. Therefore it must move out of the central plane, and when it does, its position forfeits bilateral

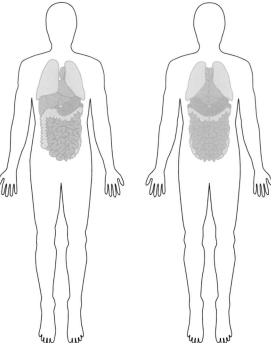

symmetry. The heart is asymmetric because its right side pumps blood around the lungs, but the left side pumps it around the entire body, so the left side is bigger. The asymmetry of the lungs is related to an asymmetry in the major airways. If you're carted off to the hospital having inhaled a peanut at a party, the odds are high that it's lodged in the right lung. If the lungs and the airways leading to them were mirror images of each other, the peanut would be equally likely to land up in either lung.

Mechanical constraints can sometimes explain why bilateral symmetry isn't feasible, but they can't of themselves explain particular choices of handedness. An intestine that bends to the right, or its mirror image bending to the left, are equally valid solutions to the problems posed by the mechanical constraints, because the laws of physics are symmetric under reflections. If the

human heart was on the left or the right, with equal probability, then we could imagine that handedness was a random consequence of mechanics. However, nearly all humans have their hearts on the left. One person in 8,500 suffers from a relatively harmless condition known as *situs inversus*, in which all internal organs are in the mirror image position. (Unexpectedly, such a person is no more or less likely to be left-handed.) A few people suffer from *isomerism*, in which one half of the normal body is doubled up with its mirror image. This can be serious, for instance they may have no spleen or two, depending which half it was.

Abnormalities like these are caused by errors in the system of genetic switches that gives developing groups of cells spatial orientation. There is evidence that in mice a gene known as Pitx2 determines "leftness" in the lungs; it also

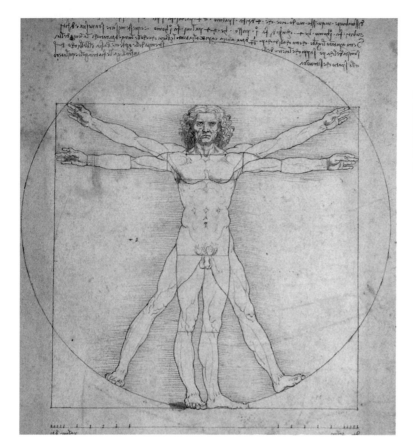

LEFT Leonardo also studied the mathematical proportions of the human body, and in particular its bilateral symmetry, to help improve the realism of his paintings and drawings.

affects the position of the heart and the development of the pituitary gland and teeth. In humans, a mutation in this gene causes Rieger syndrome, involving abnormalities in the eyes and the face.

An alternative explanation invokes the dynamics of cilia, tiny whiplike projections on cells which move rather like a millipede's legs. In the developing embryo there is a cup-shaped cavity, the node, which uses rotating cilia to direct fluid over the embryo. This flow is asymmetric, with a handedness determined by the cilia, which rotate counterclockwise. The theory is that the asymmetric flow may activate different genes on the left or right sides of the embryo, thus creating a specific handedness in all later development. This is still genetically determined as the direction of rotation of the cilia could be a necessary feature

of their own structure, which could be specified by the cell's genes.

Bilateral symmetry itself clearly goes back a long way in evolutionary history. I suspect that to begin with it was a necessary feature of the growth of living creatures, a consequence of the left-right symmetry of physics and chemistry. It turned out to be useful, because mirror symmetry gives you two of everything for the price of one. The main developments since have been minor, but highly important, variations on bilateral symmetry.

THE SYMMETRIC SNOWFLAKE

If people were snowflakes then there would be more ways for their symmetry to go wrong. Bilaterally symmetric forms have only one reflectional symmetry; while others can have more. In particular, snowflakes have several reflectional symmetries, though human psychology causes us to focus much more strongly on the sixfold rotational symmetry. In fact, any perfectly symmetric snowflake will have six distinct reflectional symmetries.

The easy way to see this is to take the simplest, most boringly regular, snowflake—a hexagon. There are six different lines that pass across a hexagon and cut it into mirror-symmetric pieces. Three of these lines join a corner to the diametrically opposite corner. The other three join the midpoint of a side to the diametrically opposite midpoint. All of them meet at the center of the hexagon, and the angles between neighboring mirror lines are exactly 30 degrees.

These same six axes of mirror symmetry exist in more elaborate snowflakes, too. For instance, the treelike dendritic shape at a corner is usually left-right symmetric, and the same shape is repeated at the other five corners.

There is something about the physics of the growing snowflake—just as there is something about the biology of the growing organism—that creates and maintains the symmetry, independently of the precise details. How can this be?

The mirror axes of the hexagon are a clue. There is a child's toy that acts in the same way: the kaleidoscope. A kaleidoscope consists of two mirrors joined at an edge and placed at a carefully chosen angle in the covering tube. You look down into the V-shape formed by these mirrors, parallel to the "hinge" where they join. At right angles to your line of sight is a random jumble of bits of transparent colored plastic, beads, paper—junk. The mirrors repeatedly reflect the junk, and every reflection introduces more symmetry. You do need to get those angles right, however. Because of a peculiarity in the geometry, odd numbers of (apparent) mirror axes work better than even ones. If you want to get sixfold symmetry in a kaleidoscope you have to make the V angle equal to 30 degrees, which is rather sharp to look into (though not impossible). Fivefold symmetry requires an angle of 72 degrees, which is nice and wide and easy to look into in your kaleidoscope. This angle is one-fifth of a full circle, so you might expect to get sixfold symmetry using an angle that is one-sixth of a circle—60 degrees. However, this actually gives

ABOVE AND BELOW LEFT The snowflake has six rotational symmetries, through multiples of 60 degrees (above left and below left). It also has six reflectional symmetries. Kaleidoscopes exploit reflectional symmetry to produce symmetric images. With two mirrors, inclined at 60 degrees, the image has three symmetries, not six (above right).

threefold symmetry (although hexagons such as snowflakes have sixfold symmetry on a 60-degree angle).

Why? Because with even numbers, various reflections of the mirrors sit on top of each other. With odd numbers, this doesn't happen. Hexagons have two different types of mirror axis— corners and midpoints of edges—but the mirror axes of a pentagon all join one corner to the midpoint of the opposite edge. That's what causes the difference.

The reflectional symmetries of a snowflake are primary. If two reflections in different mirrors are combined, the result is a rotation. For instance, if we reflect an object in a mirror and then reflect the result in a mirror placed at 30 degrees to the first, the result is a 60 degrees rotation. The angle doubles. In contrast to this, no combination of rotations in the plane give a reflection.

The most important feature of a kaleidoscope is that its overall effect is a highly symmetric image, far prettier than the random junk from which it is created. And it works whatever junk you use. It's not hard to see why. The junk provides detail and texture, but not the overall structure. The mirrors create the overall structure—symmetry—and the symmetry is the same, whatever the junk might be. So we can dissect the kaleidoscope pattern into two independent parts—one creating the detail, the other the symmetry.

If we could do the same for the physics of a snowflake, we would have a really satisfying answer to our central question. But what plays the role of the kaleidoscope? What plays the role of the plastic junk?

SYMMETRY OF PHYSICAL LAW

The snowflake's "kaleidoscope" must surely be due to the laws of physics, because they are the ultimate source of all of nature's symmetries. It was Albert Einstein, above all others, who realized the importance of symmetry in the laws of physics. Earlier mathematicians had discovered a link between symmetry and "conservation laws," which tell us that certain physical quantities such as energy or momentum can neither be created nor destroyed. Einstein, however, went further and made symmetry the foundation of all of the deep laws of nature.

BELOW Visible traces of the quantum world: tracks of fundamental particles in a physical experiment. The spiral paths, created by magnetic fields reveal the electrical charges of the particles.

In essence, his view was that physics must act in the same way at all points of space and at all instants of time. No place, and no time, should be special. Yes, different events can happen in different places or at different times, but the laws that govern these events are identical.

In Einstein's hands, this principle led to the theories of special and general relativity, which still provide the basis for today's understanding of mechanics, electromagnetism, and gravity. The other great revolution in physics, quantum theory, also hinges on the principles of symmetry—but these are often esoteric and difficult to comprehend without a big dose of advanced mathematics. Many of the symmetry principles of quantum mechanics are about the features of fundamental particles. Basically, they tell us that the laws stay the same when the particles are replaced by different, related particles. In other words, particles are not individuals, but members of symmetrically related families.

Reflectional symmetries make modern physics interesting but also bedevil our attempts to reduce the universe to simple, elegant laws. Other symmetries, such as translations and rotations, cause little trouble. If an elephant, or an electron, is moved through space, or rotated, it still behaves just like an elephant or an electron would. If an elephant or an electron is moved through time —say from yesterday at noon to tomorrow evening—it still behaves just like an elephant or an electron would. This is hardly surprising because it is possible to perform the motion—in the case of time, just by waiting long enough.

Going backward in time, which is the temporal analog of a reflection, is another matter. If we could observe an elephant or an electron in backward time, would it still obey the same physical laws? This time we can't do the experiment, but we can examine the mathematical structure of the laws—and what they revealed, at least until recently, was symmetry under time-reflection. The same went for symmetry under ordinary, spatial mirror-reflection. Finally, there was a quantum-mechanical reflectional symmetry known as parity. Parity "reflects" electrical charge—it turns positive to negative and negative to positive.

I use the past tense because, although most of physics has these three reflectional symmetries, we now know that certain subatomic particles behave in a manner that is inconsistent with either mirror symmetry or parity symmetry. The fundamental forces of nature come in four kinds—gravity, electromagnetism, the strong nuclear force, and the weak nuclear force. The first three forces are symmetric for mirror symmetry, time-reflection, and parity. But for some reason the weak force is not. As Wolfgang Pauli, one of the 20th century's leading physicists, put it: "God is a weak left-hander."

This discovery raises an intriguing possibility. Just as there are functional people whose organs are mirror images of the usual ones, so there might be another perfectly functional universe whose laws are the mirror image of ours—one in which God is a weak right-hander. Indeed, it looks as if our universe started out left-right symmetric at the time of the Big Bang, but then deviated into its current left-handed state.

Did a right-handed universe split off from ours at the beginning of time?

Still more recent developments introduce yet another sort of reflection into what many physicists think ought to be the laws of nature, known as supersymmetry. Every known fundamental particle is thought to possess a supersymmetric partner, known as a sparticle. The electron, for example pairs with the selectron, and the quark with the squark. Could there be a ghostly partner universe, made from sparticles, interacting with our own?

THE ASYMMETRIC BRAIN

Some important asymmetries show up most clearly when we look not at form but at function. The two sides of our brains look very similar, but they carry out rather different tasks. The old story was that the right side is more strongly involved in visual perception, spatial orientation, and the recognition of faces and objects; while the left is used more for language, the programming of complex sequences of movement, and awareness of our own bodies. It now looks as if this is far too simplistic—the difference is not in subject matter but processing style. Through the use of scanning equipment we can now observe which parts of the brain are active while it is thinking, and all mental faculties seem to be shared by both halves. But, the left half goes for details and the right for the broad picture.

The difference is puzzling. Structurally the brain appears symmetric, composed of two roughly identical hemispheres of nerve tissue. But on closer examination, the two halves differ in a number of ways; even the pattern of folding of the tissue is different. The functional asymmetry of the brain is much more marked than the structural asymmetry, however, and it is related to handedness, though not in a clear-cut manner. For example the left hemisphere of the brain is dominant for language in over 99 percent of right-handed people, but only 60 percent of left-handed or ambidextrous people. In about 30 percent of ambidextrous people, the right hemisphere is dominant for language, and in the remaining 10 percent neither hemisphere is language dominant. To complicate the picture, the right hemisphere has more linguistic capability than was previously thought. Why are our brains like this? Perhaps for the same reason that our intestines break symmetry—there isn't room to be symmetric. Both our linguistic and

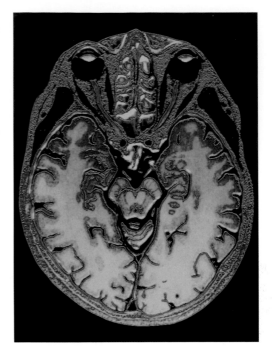

visual abilities require huge amounts of information processing, and it may be that the only way to include both in the same human-sized brain is to specialize the powers of the two halves.

Animals seem to have a curious affinity for symmetry, and this results in their placing a lot of importance on dualities like left/right. This affinity probably comes from the evolution of the brain. Animals' ability to recognize objects is not wired in, although the ability to learn seems to be. The brain is a neural net, complex circuits of nerve fibers, and the training of such circuits can introduce unexpected symmetry effects. For example in some bird species females prefer to mate with males that possess symmetric tails. There are two competing theories to explain this, both of which may be right. One is sexual selection. A bird needs good genes and a properly functioning developmental system to

LEFT Superficially, the two halves of our brains look very similar in shape and size. However, they contribute very different things to the workings of the brain. The main difference is not subject matter, as used to be thought, but how perceptions are processed.

BELOW A computer analysis of the most significant features of the human face shows that the average face (center) is gender neutral. The strongest difference in faces is that between male (left) and female (right). Until recently, it was difficult to train a computer to distinguish between the sexes. As a result of this kind of

analysis, it has become very simple. The important differences cannot be detected by looking at any single feature, such as the nose or mouth, on its own. Instead, it is necessary to look at coordinated changes to all of the features simultaneously.

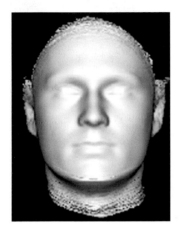 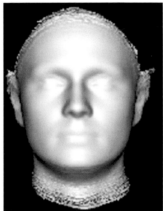 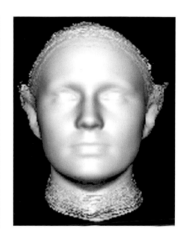

ensure that its tail is symmetric, so females that prefer to mate with symmetrical males will produce better offspring, thereby perpetuating the preference for symmetry. The other theory is that the preference for symmetry is an accidental by-product of a visual system that has been trained to recognize tails. The bird's visual neural net must respond strongly when its eyes see a tail, but tails come in a variety of shapes, and for every tail that is lopsided on the left there is probably one that is similarly lopsided on the right. So any neural network that responds strongly to both of these should respond even more strongly to a symmetric tail because it resembles both lopsided versions and so counts twice when generating the "this is a tail" response.

Humans have a similar preference for symmetry or near symmetry, and it may have arisen for similar reasons. Recent research

indicates that women have more or deeper orgasms if they mate with males whose faces are nearly symmetric. Is this sexual selection in action, then? Possibly. But as with birds, our love of symmetry may be just another accident of neural net architecture. Or the two may be mixed up.

In 1996 cognitive scientist Alice O'Toole and computer scientist Thomas Vetter carried out a computer analysis of images of faces. They first found and analysed the average face. Any given face can then be compared with this average by subtracting the average face and seeing what patterns may be found as a result. Our visual senses are sensitive to the "correction term" to the average face, the variable that contains the most information about how that face differs from the average. And what is that variable? It represents the difference between typical male and female faces.

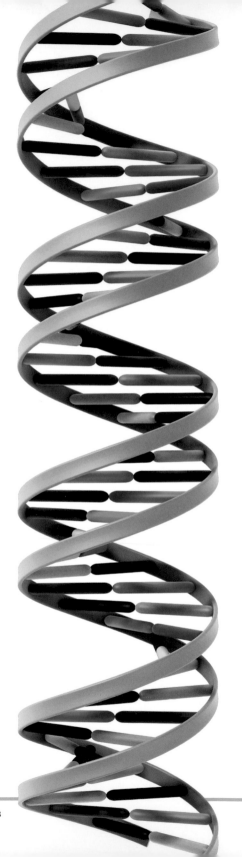

PHYSICS & GENES

The puzzle of life's intrinsic handedness continues at the molecular level. The mirror symmetry of the laws of physics leads to fascinating features of molecules. Some molecules are bilaterally symmetric, but most are not. Any of the latter can exist in two distinct forms, each the mirror image of the other. Their physical and chemical properties will be the same (though if one reacts to light by rotating it to the left, its mirror image will rotate to the right). Their biochemical properties, however, can be very different. Not because of an asymmetry in physics, but because of an asymmetry in biology. Living creatures are built out of asymmetric molecules, and on this planet, at least, a definite handedness is preferred.

This is true of both DNA and the proteins that it specifies. If you try to eat food based on mirror-image proteins, you obtain very little nourishment. A left-handed molecule may taste different from its right-handed mirror image because our sense of taste has biased handedness. And the DNA double helix coils with a definite handedness—it is right-handed, like a common corkscrew. It could equally well have coiled the other way—everything would have worked just as well—but on this planet it doesn't. (There is, however, a rare form of DNA known as Z-DNA, that forms a left-handed coil out of the same components as standard DNA, although only in special circumstances.)

LEFT The laws of physics are (with one rare class of exception) symmetric under reflections. As a result, chemical molecules that are not themselves bilaterally symmetric can exist in two different forms, each a mirror image of the other. The DNA molecule is a double helix with a specific "handedness." Its mirror image is equally feasible in chemical terms, but is not found in any of the Earth's life-forms, for a number of biological and evolutionary reasons.

5 *MIRROR SYMMETRY*

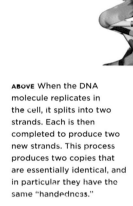

It's reasonably easy to explain why a mixture of coiling directions within a given species isn't a good idea, at least for sexually reproducing organisms. Life on Earth falls into two big groups—prokaryotes (mainly bacteria) and eukaryotes (nearly everything else). Eukaryotes are built from one or more cells, and their genetic material is packaged in chromosomes. In each chromosome, other than the sex chromosomes, there are two sets of DNA, one from the father and one from the mother. These lengths of DNA cross over every so often, with father and mother switching places. This process, called recombination, shuffles the organism's genes. The crossover mechanism wouldn't work very well if it had to join left-handed DNA to right-handed DNA.

All very well, but this doesn't explain why organisms from different species all have DNA with the same handedness. The explanation here may well be evolutionary. When DNA replicates, its handedness carries over automatically to the copies. If we are all descended from a single "founding" life-form—which seems likely because of the universality of genetic mechanisms—then we've all inherited the handedness of that life-form's DNA. If it had been the mirror image instead, we'd all have our DNA the other way around. So the handedness of DNA may be a "frozen accident." This theory implies that if we ever encounter an alien race from another world whose biochemistry is also based on DNA, then there's a fifty-fifty chance that its DNA will coil the opposite way to ours.

Alternatively, the direction of DNA coiling, and the handedness of proteins, may have arisen because our universe is biased. To be specific, the handedness of biological molecules may be an evolutionary consequence of the asymmetry of the weak nuclear force. Remember, there are four forces in nature, and the weak force is the only one whose behavior does not stay the same

ABOVE When the DNA molecule replicates in the cell, it splits into two strands. Each is then completed to produce two new strands. This process produces two copies that are essentially identical, and in particular they have the same "handedness."

For similar reasons, when two DNA strands—one from each parent—recombine in sexually reproducing species, both sets of DNA must have the same handedness. The two strands of the helix can be pulled apart to make copies of the genetic information.

when the universe is reflected in a mirror.

One consequence of this asymmetry is that the energy of a molecule and that of its mirror image are not quite the same. Until recently this was thought to be unimportant because the difference is vanishingly small—to be precise, one part in a hundred quintillion. About 25 years ago, however, the physicist Dilip Kondepudi showed that if nature is biased in favor of the lower energy version of some biologically significant molecule (even by this tiny amount), then within a mere hundred thousand years a massive 98 percent of those molecules will be of the lower energy variety. The difference is amplified by the reproductive processes of life.

6

ROTATIONAL SYMMETRY

Rotation through zero degrees is important to mathematicians, but all it means is that a snowflake, which has the six rotational symmetries 60, 120, 180, 240, and 300 degrees, looks exactly the same if you do absolutely nothing to it. Replace "snowflake" by any other form, and the same goes—big deal. But if you leave out this "trivial" symmetry— "do nothing"—then the mathematics gets in a terrible tangle. It's like trying to do arithmetic without zero.

The snowflake has discrete rotational symmetry—only certain specific angles do the trick. Some shapes are symmetric under any rotation whatsoever—they have continuous rotational symmetry—and the archetype shape here is the circle. To the Greeks, the circle was the perfect form. Every point was at exactly the same distance from one special point—the center. This, of course, is why you can draw a circle with a pair of compasses. Stick the sharp end into the paper so that it can't move—that's the center. Swing the other end, usually a pencil point, around in a sweeping arc—that's the circle. Keep the compasses opened to a fixed distance, and the distance from center to pencil mark never changes. That's why the Greeks were so enamored of circles—with every point bearing exactly the same relationship to the whole as the others do.

When a pebble is thrown into the smooth unruffled surface of a pond and hits the water, it creates an intricate pattern of waves—and all of them are circles, or parts of circles. Why? If it's a small pebble, the disturbance is, near enough, confined to a single point. Since water is fluid, though, the disturbance jostles neighboring

regions and spreads. If there is nothing to distinguish one direction from another—as for an unruffled surface—the disturbance spreads equally fast in all directions. At any instant the distance it has traveled from the pebble is the same, whatever the direction. In short, the ripples made by the pebble are circles.

The circles expand, and because that first disturbance causes the water to oscillate up and down several times, we see several circles, all with the same point at the center. Later circles are fainter than the first one, because the oscillating ripples aren't quite so big.

When the ripples hit the edge of the pond, they bounce, like light reflected in a mirror.

LEFT AND BELOW Ripples on a flat pond have circular symmetry because they spread at equal rates in all directions (left). A splash also starts with circular symmetry, but as the liquid rises this symmetry is destroyed, leading to a less symmetric, but still highly patterned, crown shape, with a series of equally spaced spikes (below).

When one ripple meets another, they pass through each other unchanged, except for where they intersect. Here, the two ripples join forces temporarily. When peak meets peak we see a higher peak; when trough meets trough we see a deeper trough. When peak meets trough, the ripples cancel each other out.

What if a raindrop—rather than a pebble—hits a pond?

Again, we see circular ripples (a familiar sign that it is raining). But at the very center, something more dramatic happens. A spherical raindrop hits a flat surface and it splashes. For an instant, a circular depression appears, then the water leaps skyward to form a steep, circular wall. (All this is tiny, not much bigger than a raindrop—you need special photographic equipment to freeze time and capture its transient beauty.) The wall grows, and then the top of the wall develops ripples. A series of spikes—sharp jets of water—spurts from the top of that circular wall; at the tip of every jet is a rounded droplet.

For all the world, the splash looks just like a crown. The jets are more or less equally spaced—so the splash, like a snowflake, has discrete rotational symmetry. Why doesn't a splash keep the full circular symmetry of the drop that caused it? Is there any relation to the sixfold symmetry of a snowflake?

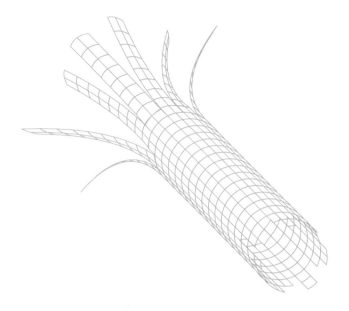

LEFT AND BELOW Volvox (below) is spherically symmetric, and it keeps its smaller young inside itself until they are ready to be released. The tubulin molecule (left), an even tinier miracle of biological engineering, is a cylinder. It grows by adding units one at a time, but can shrink ten times faster by splitting along its length. Amoebas move by building and destroying tubulin.

MICROSCOPIC MARVELS

A fast camera can freeze a splash—new instrumentation often lies behind major scientific advances. The 15th century saw the invention of one of the most important new instruments ever. It began with powerful single-lens magnifying glasses, and by 1674 the Dutch naturalist Anton van Leeuwenhoek had improved them so much that he could see individual bacteria. Others introduced combinations of several lenses, and the microscope was born. Science has never been the same since.

The microscope revealed a hidden world of astonishing diversity and activity. A single drop of pond water contained more life-forms than the naked eye could see in an entire field. Biology took an enormous leap forward.

Microscopic organisms are far simpler than cows and caterpillars, but they are still amazingly complex—especially when you focus

not on their forms but on how they work. An amoeba may look like nothing more than a formless blob of jelly, yet it moves through its microcosmic world with a convincing air of purpose—or so it seems.

The microscopic world is a realm of beauty and surprise. The slipper-shaped Paramecium has a fringe of tiny hairlike cilia, whose ripples miraculously cause it to move through its watery habitat. This is symmetry, of a kind, the space-time symmetry of a traveling wave, like the millipede's legs (see pp. 40–41). Some microscopic organisms are symmetric in a more literal sense. The alga Volvox is a spherical network of small green dots, each sporting two cilia. Inside it are contained the smaller green spheres—the next generation. And inside them, amazingly already in existence, can be found the generation after that. Some species of diatom—single-celled plants with a silica shell—have the same combination of reflectional and rotational symmetries as a snowflake and a starfish, the symmetries most commonly associated with a regular polygon.

Asymmetric as the amoeba may be, its purposeful motion is created by a mix of symmetry, randomness, and dynamics. An amoeba is a cell, a single-celled organism. A cell is a complex piece of chemical machinery encased in a membrane, and it has genetic equipment which it keeps inside its nucleus when not actively in use. The genes come into their own when a growing amoeba divides into a pair of smaller cells, each a new amoeba in its own right. Amoebas multiply by dividing.

Genes do a lot of things, most of which we don't understand, but the one that we do understand is that they make proteins. Like most cells, the amoeba has a skeleton, a network of hollow rodlike structures known as microtubules. They are made from a protein called tubulin, which comes in two slightly different forms,

alpha and beta. Somewhere in the amoeba's genes are recipes for alpha- and beta-tubulin. But there are no genetic recipes for what these molecules do, their behavior is governed by the laws of physics.

One thing these laws make them do is assemble themselves into a tube, similar to a rolled-up checkerboard whose black squares are alpha-tubulin and whose white squares are beta-tubulin. The result is called a microtubule. Amoebas move by tearing down their microtubular scaffolding and rebuilding it elsewhere. They build it by adding successive rings of protein to the end of the tube. They destroy it by splitting it lengthways like peeling a banana. Both processes are dynamic, both are symmetric…and yet they are different. Construction is ten times slower than destruction.

Tubulin is a key constituent of the cells of all living creatures, not just the amoeba. It is extruded in quantity by a strange item of molecular machinery, the centrosome. The centrosome itself is made from tubulin. It consists of two identical bodies, centrioles, which sit at right angles to each other. Each centriole is a twisted cylinder made from 27 microtubules grouped in nine sets of three, with perfect ninefold rotational symmetry— but no reflections. No one knows why. Very little is known about the detailed workings of this diminutive machine, but we do know that it plays a key role in cell division. Cells pull themselves and their chromosomes apart with microtubule cables.

First the centrosome replicates, then it extrudes microtubules; as the cell separates into two pieces, these cables latch on to the duplicated chromosomes and pull one set into each daughter cell. So at the heart of life's reproductive machinery we find an elegant but enigmatic symmetry.

SYMMETRY UNDER THE SEA

Rotational symmetry, also called radial symmetry, is common in living creatures, and it usually goes hand in hand with reflectional symmetry. Mathematically, it is possible for an object to have only rotational symmetries. A classic example is the Isle of Man's national symbol, a set of three running legs that is symmetric under rotation through multiples of 120 degrees, but not under reflections. Some viruses have fivefold rotational symmetries but no reflectional ones.

On land, rotational symmetry is most often seen in flowers, but under the sea it is well represented among animals such as sea anemones, corals, and—of course—starfish. The overall form of a typical starfish has fivefold symmetry—rotation through multiples of 72 degrees, rather like a five-pointed snowflake. Like the snowflake, the starfish is also symmetric under reflections, one for each of its arms.

Starfish are members of the class (more technically, the phylum) of creatures known as echinoderms, which also includes sea lilies, sea urchins, sea cucumbers, brittle stars, and sea daisies. The prevalence of flower names or the word "star" in their names reflects these animals' visually most striking attribute—rotational symmetry. They are built on a radial body plan rather than the more familiar bilateral one. The radial geometry is a result of the creatures' development, but it sets in at a surprisingly late stage. Echinoderms pass from a roughly spherical egg to a larval stage that typically is bilaterally symmetric—not unlike vertebrate embryos and indeed many invertebrates. Only later is the entire body plan rearranged by a complicated metamorphosis into the rotationally symmetric form. The underlying bilateral symmetry gives way to a radial structure, dominated by five water-vascular

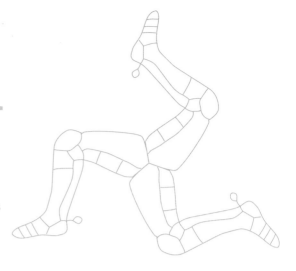

ABOVE An intriguing shape with threefold rotational symmetry, but without reflectional symmetries, is the Isle of Man's "running legs" symbol. Any two of the legs together look like those of a running man, but there are only three legs altogether, each arranged at angles of 120 degrees.

canals. This network of vessels is filled with a watery fluid and acts as a hydraulic system to move the animals' tube feet.

The fivefold structure is obvious in common starfish species. It is less obvious in sea urchins, which look as if they have their future evolutionary sights set on spherical—or at least polyhedral—symmetry, but it is easily seen in their skeletal structure. This is based on a roughly spherical shell split into five parts, arranged like the segments of an orange about a central axis. The sand dollar, a flat pentagonal object found on many beaches, is the internal skeleton ("test") of a flattened type of sea urchin known as a cake urchin.

Possibly the evolutionary history of echinoderms began with a bilaterally symmetric form, slowly modified to have strong elements of fivefold symmetry; then later the fivefold symmetry pretty much obliterated the original bilateral symmetry. Or perhaps they followed their current developmental route from very early on. It is not known why these creatures have such strong radial symmetry, but it is thought that fivefold symmetry offers

advantages of structural strength compared to, say, threefold symmetry.

Not all starfish have a fivefold body plan, though. Species with 7 arms are relatively common, one deep-sea family of starfish has between 6 and 20 arms, and in the Antarctic there are species with up to 50 arms. A common species of sun star, found in northern waters, has 10 arms, and the spiny sun star can have 15. No particular structural reason—akin to the Fibonacci numerology in flowers and its origin in the dynamics of plant growth *(see pp. 126–127)*— is known to explain any of these numbers.

The symmetry of echinoderms sometimes carries over to their motion, in the same way that the symmetry of centipedes and millipedes leads to patterns in their movements. A striking example occurs in feather stars (crinoids) which swim by thrashing their arms up and down.

In a ten-armed species, for example arms 1, 3, 5, 7, 9 will thrash upward while arms 2, 4, 6, 8, 10 thrash downward; then the pattern reverses. Mathematically speaking, this is a typical dynamical pattern for a tenfold symmetric system.

Of course the detailed structure of a starfish is not perfectly symmetric under rotations, just as that of a human being is not perfectly symmetric under left-right reflection. But it is the nearness to perfect symmetry, rather than the departure from it, that most requires explanation. If just one arm of a starfish includes biologically special organs, then technically the creature becomes bilaterally symmetric— but this is no ordinary bilateral symmetry. It is slight bilateral departure from a near-perfect fivefold symmetry, and that is what requires explanation.

PATTERNS IN LIGHT

Rotational symmetry can often be seen in physical patterns, as well as in biological ones. In fact, there are many physical patterns with complete circular symmetry—unchanged by any rotation, through whatever angle. One of the most familiar examples is the rainbow.

Most explanations of the rainbow concentrate on its colors, but they are only part of the puzzle. When light rays pass from one medium to another—such as from air to water—they bend, an effect called refraction. Isaac Newton demonstrated how refraction can split white light into "all the colors of the rainbow" by passing sunlight through a slit in a blind and then through a glass prism. On the far side of the prism he saw colored bands, in the usual "rainbow" order: red, orange, yellow, green, blue, purple. Light is a wave and its color depends on its wavelength; light rays of different wavelengths bend through different angles. The same effect also creates the colors of the rainbow. Each drop of rain acts like a tiny prism and splits the Sun's white light into its constituent colors.

This explains why rainbows have many colors, but a more interesting mathematical question is the rainbow's shape. Each colored bow is an arc of a circle, as we'll shortly see. Other geometric features of rainbows also demand explanation. Sometimes—especially if there is a lot of rain—we can see two rainbows. The colors run in reverse order in the second rainbow, and the sky between the two is quite dark. Why?

Let's think about what happens when sunlight encounters a drop of rain. Think of light rays of just one color—take the color red as an example. When a bundle of parallel red rays from the sun meets a raindrop, passing from air into water, it is first refracted—it undergoes a sudden change

BELOW When sunlight passes through a prism it splits into its component colors because each color of light bends through a slightly different angle.

of direction. Then it hits the far side of the drop and is reflected—it bounces off the surface. Finally it is refracted again as it emerges from the water drop back into air.

These changes of direction concentrate the light rays so that they turn through a particular "critical angle." Because the whole setup is symmetric under rotations about the axis joining the raindrop to the distant Sun, the returning rays sweep out a bright cone. When we look at rain with the Sun behind us, light is sent back toward us along all of these cones—one cone of each individual color for each individual raindrop. Our eyes receive light from a drop only when the eye lies on one of these cones. Because the cones have a circular cross section, each color therefore forms a circular arc. Because light of different colors is refracted through different angles, each colored arc has a slightly different radius, hence the layers of color in a rainbow.

Sun

Raindrop

LEFT When sunlight encounters falling raindrops, each drop acts like a more complex prism. Not only is the light split into different colors, but each color is also concentrated along a specific direction. When the Sun is behind us and the rain ahead, we see a band of colored arcs—a rainbow. Each arc is formed by those droplets that emit light in just the right direction to meet our eyes.

When there are two rainbows, the second one is formed by light that is reflected twice inside the raindrop instead of once, and this also reverses the order of the colors you can see. In principle this always happens but the second rainbow may be too faint to be seen clearly. The sky between the two rainbows is dark because very little of the light that hits a raindrop is scattered in directions that lie between the two corresponding cones.

There are other optical effects with similar explanations to the rainbow. With ice crystals taking the place of raindrops and the Moon instead of the Sun, an arrangement of different arcs centering around the Moon can be created, known as a halo. You can often see a halo around a full moon on a frosty night. The most amazing of such effects, though, is the glory.

If you stand with the Sun behind you and look into mist or fog, or sometimes just murky water, you can see a dark shadow of yourself with a bright multicolored halo about your head. Miraculously, if other people are standing next to you their heads appear not to have these halos. However, each of them sees a halo about his head, but not yours, and not his other companions' heads. This almost mystical effect occurs because light from the Sun bounces around each droplet of mist, bouncing several times, and then travels along the surface of the droplet for a short distance, before returning straight back the way it came. In order for a ray of light to enter your eye, it must have passed very close to your eye on the way in—that is, around the edge of your head. So you see the halo around your head's shadow. Only your head is in the right place for the halo of glory to be seen by your eyes, other people's heads are too far from your eye for their halo of glory to be seen.

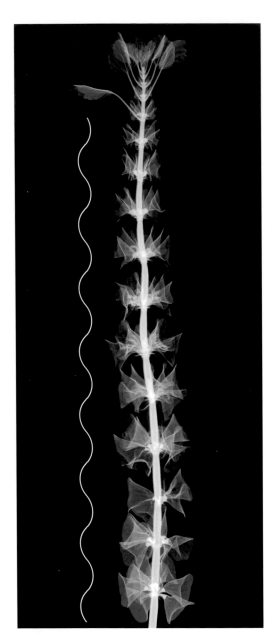

ABOVE A feature of plant growth is for certain parts, such as branches, to repeat the same structure many times.

TALE OF THE TOADFLAX

I've mentioned the shape of flowers several times, but until now I haven't examined them in detail. They have the same kind of radial, or rotational, symmetry as the snowflake and, like the snowflake, their shape is a record of how they grew. However, the growth of a snowflake is a physical process, while flowers are biological. As always, this creates substantial differences because of possible genetic influences. Some recent work sheds a fascinating light on the role of genes in plant symmetry.

Rotational symmetry is possible in flowers because they originally form and then grow at the tip of a roughly cylindrical shoot. A cylinder has circular symmetry, and identical objects arranged around a circle are therefore quite likely to have some relic of that symmetry. The mathematics of growth provides a catalog of possible patterns; genetics can choose among the entries in the catalog.

Plant genetics has been widely studied in the laboratory, but it has never been clear how relevant those studies are to plants that grow in the wild. In 1999 botanists Pilar Cubas, Coral Vincent, and Enrico Coen made a close genetic study of a plant originally described in 1749 by the Swedish naturalist Linnaeus, the toadflax *Linaria vulgaris*. Mature wild toadflax plants have five petals but only bilateral symmetry. The two uppermost petals poke up like a pair of rabbit ears, the ones either side of them hang down like jowls, and the lowest pokes out from between the jowls like a tongue with a long spike attached to it (which contains nectar).

Linnaeus noticed that sometimes a mutant form of the toadflax occurred, one with fivefold radial symmetry. And instead of the three clearly different types of petal found in the normal variant, in the mutant form all five petals bore spikes. The difference is a consequence of the

6 ROTATIONAL SYMMETRY

development of the petals and the other parts of the flower. Roughly speaking, both of these forms have the same components each arranged in roughly the same positions, however, in the common toadflax different petals develop in different ways, while in the rarer mutant they all develop in the same way.

Cubas looked for genetic changes that would explain these developmental differences and traced them to the Lcyc gene, an identical gene that controls asymmetry in another plant, the antirrhinum (snapdragon). What they were expecting to find was some small mutation in the DNA sequence on the Lcyc gene, some tiny typographical error in the DNA code, which would make it behave differently in the two cases. What they actually found was much more interesting than this. There is no difference in the DNA code itself. Instead, there is an epigenetic difference, a chemical marker that can be transmitted from parent to offspring but is not copied by the normal DNA replication mechanism.

This difference is a phenomenon called methylation. It has been known for some time that methyl molecules are often attached to segments of an organism's DNA. They are not part of the organism's DNA code—instead, they label the segment so that it has a different effect. In this case, the label stops the Lcyc gene being acted on, or expressed, during the development of the mutant variety. Methylation need not always have this effect. Sometimes it ensures that genes are expressed, sometimes the opposite—it all depends on which gene and in which context. In this case, though, it essentially attaches a chemical note to the Lcyc gene saying "ignore this." So when the developmental process gets to the stage at which it would normally activate the Lcyc gene, build the protein that the gene specifies, and use it to make the petals different, it doesn't do that. It plows ahead with

TOP AND ABOVE The presence or absence of a chemical marker can cause changes in plant shapes. This happens in toadflax (top).

Differences between normal toadflax (above left) and mutant toadflax (above right) can be observed in flower diagrams.

its default process, which is to keep them all pretty much the same.

There are several points to this tale. One is that there is more to life than just its DNA code; many other processes play a part in making an organism. Another is that biology is rather good at adapting patterns from the mathematical catalog. Here it takes what might otherwise be a five-pointed star-shaped flower and subverts its development with a dash of Lcyc to create an attractive platform for pollinating insects to land on. The third is that methylation is an extremely rare kind of mutation in the laboratory— yet here it is in the first example of a wild mutation ever studied. I leave you to draw your own conclusions.

WANDERING SPHERES

Rotational symmetries come into their own in the heavens. To the Greeks, the planets were specks of light in the sky, distinguished from the "fixed" stars only by their insistence on wandering around. Later, it became apparent that each planet is an entire world, similar in some respects to our own, but mostly different. Most planets have an atmosphere; only Mercury has none worthy of the name. Planetary atmospheres contain very different proportions of gases from ours. Jupiter's is dominated by hydrogen and helium, for example.

In most respects, each planet is unique. Mercury is an atmosphereless, crater-strewn world of smashed rock. Venus is an acid hothouse riddled with erupting volcanoes. Mars is a frozen desert. Jupiter is a striped giant sporting a large circulating Red Spot. Saturn has spectacular rings of floating rock. Uranus is a featureless globe of cloud. Neptune is spinning on its back. And Pluto is plain weird. It was recently redefined to be a dwarf planet, not a true planet. Images from NASA's New Horizons mission show

that it has a very complex, strange surface of frozen gases and ice.

Individual though its members may be, our solar system's retinue of planets also conforms to some common patterns. Jupiter, Uranus, and Neptune also have rings, Venus and Mars also have craters, Saturn's atmosphere also has stripes. And some of the differences have common explanations—Einstein's point that the laws of physics ought to be the same everywhere, even if they produce different consequences in different places. The planetary atmospheres, for instance, are mainly determined by which gas molecules their gravitational fields can prevent from escaping. More massive planets can hold lighter gases, and that's exactly what we find.

An obvious common pattern is the planets' shapes and movements. They are all round and they all spin about some axis. This means that to a first approximation the physics of planetary surfaces, atmospheres, and internal structure fits into a common frame—the things we should expect to see in a rotating, spherically symmetric system. Which raises the question: what should we expect to see?

6 ROTATIONAL SYMMETRY

LEFT TO RIGHT: PLUTO, JUPITER, EUROPA, AND SATURN The planets and moons of the solar system are individuals, varying wildly in their surface colors and textures. For example, the moon Europa is characterized by its icy surface and long, intersecting cracks. However, there are also many common features. All of the planets, and the larger moons, are spherical in shape. Their rotation typically creates features with circular symmetry, such as Jupiter's cloud bands and Saturn's beautiful rings.

Spherical symmetry alone would lead us to expect very little—just a bland, featureless ball, like the superficial appearance of Uranus. However, a planet's rotation can destroy spherical symmetry, reducing it to circular symmetry about the axis of rotation. So we might expect the main planetary features to be symmetric under all rotations about the planet's axis, meaning that everything comes in circular bands.

Saturn's rings are like this, subject to some fine details such as braided rings and slightly noncircular rings. Seen from either of the planet's poles, nearly all of the rings form a series of circles and circular bands, separated by occasional circular gaps, all lying in the same plane as Saturn's equator. Jupiter's striped atmosphere is also consistent with circular symmetry. Seen from either of the poles, its colored bands of cloud form circles, centered on the pole. If you sandwich a layer of fluid between a solid sphere and a slightly larger transparent spherical shell, and spin the whole thing, the fluid spontaneously arranges itself in bands, a bit like Jupiter's. The reason in this case

is the effect of the forces generated by the rotation, which cause the fluid to circulate; the rotational symmetry leads to a preference for banded patterns. On Jupiter, these forces are also acting—together with the effects of heat. The outer layers receive some heat from the Sun but not much, so it is still very cold. The deep layers are far hotter. This heat difference also contributes to the striped patterns.

However, all this is a bit too convenient. Some structures on circularly symmetric planets are not circularly symmetric. A glaring example is Jupiter's Great Red Spot, an oval region nearly as big as the entire surface area of the Earth, which has been there for over 300 years. Moreover, it moves relative to the rest of the planet. It is, perhaps, a kind of perpetual Jovian hurricane. Whatever it is, it does not have rotational symmetry about Jupiter's axis. Nevertheless, experiments with rotating fluids indicate that single giant vortices of this type are typical of rotationally symmetric systems. All this seems to teach us two lessons: symmetric systems often behave symmetrically, and sometimes they don't. The puzzle deepens.

7

TILING PATTERNS

Let us not forget the original purpose of all these musings about mathematical regularities in nature—to track down the elusive patterns of snowflakes. What we've understood so far is that whenever you see a pattern, a symmetry can not be far away. Or several. These symmetries ultimately come from regularities in the laws of nature. They reflect the fact that ours is a mass-produced universe, made from myriad copies of identical components.

When Kepler performed his thought experiments about snowflakes *(see pp. 12–13)*, he ended by speculating that their sixfold form must be related to the crystalline nature of ice and that therefore crystals also must be made from myriad copies of identical components. "The formative faculty of Earth does not take to her heart only one shape; she knows and is practiced in the whole of geometry. I have seen at the Royal Palace at Dresden, in the Stables, a panel inlaid with silver; from it a dodecahedron, the size of a small nut, projected to half its depth, as if in flower."

ABOVE AND RIGHT Crystals often show geometric regularities, with surfaces composed of many flat planes, meeting each other at specific angles (above). The visible regularities of crystals are clues to their deeper physical properties. In fact, they are signs that the atomic structure of a crystal is that of a regular lattice, with numerous symmetries. The lattice of graphite, a form of carbon, is made from parallel honeycombs, with one carbon atom at each corner (right). These planes can easily slide over each other, which is why graphite is soft. Diamond, another form of carbon crystal, has a different lattice structure to graphite and is extremely hard.

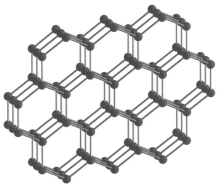

Crystals are well known for the remarkable beauty of their patterns. Quite simply, they look mathematical. Salt crystals, for instance, are tiny cubes, and in the laboratory, or even at home, you can create big cubes by growing a crystal in a strong solution of salt in water. Almandine crystals, a form of garnet, are purplish brown dodecahedra—not the dodecahedron of the Pythagoreans, with regular pentagons for faces, but a less regular form known as a rhombic dodecahedron, with rhombuses for faces. Still mathematical, just slightly less well known. Gypsum forms long prismatic structures like cut glass; cassiterite—crystalline tin—makes glittering pyramids; magnetite comes as shiny black octahedra.

In the early days of crystallography, however, none of this was clear. Innumerable irregularities distorted the clean polyhedral forms that we now associate with crystals. Fluorite, for example can form individual cubes, but as often as not several cubes grow through each other, at funny angles, a phenomenon known as twinning. In this case the twinned crystal has less symmetry than one growing on its own without interruption, but sometimes it works the other way around and several individuals join forces to create a shape with more symmetry than normal. Crystallographers call this phenomenon pseudosymmetry.

When you know the mathematical rules that underlie crystals, then it's not too hard to puzzle out what's what; but when your task is to deduce the rules from the shapes you find, twins and pseudosymmetries are confusing.

Confusions notwithstanding, we now have a beautiful mathematical theory of crystals, and its central ingredient is symmetry. Not the symmetry of the lump of mineral that a geologist digs up, but the symmetry of its submicroscopic structure, the arrangement of its atoms. It is worth remembering that although the concept of atoms originated with the ancient Greeks, it is only within the last century or so that the scientific world has been convinced that all ordinary matter is made from countless tiny atoms. Kepler did not know of the atomic basis of matter, but he may have been aware of the Greek speculations. Certainly his whole discussion of the shape of snowflakes quickly homes in on the idea that they are made from large numbers of identical tiny units, fitted together in regular patterns. He calls them globules, not atoms, though.

Our task, like Kepler's, thus divides into several separate ones. One is to understand the regularities of crystals and how the symmetries of their atomic structure affect their large-scale form. Another is to understand how crystals grow, because their final form depends on their past history.

We also need to understand what ice is, how it forms, and why it is so different from liquid water; and we need to find out how the conditions under which ice crystallizes affect the shape of the resulting crystals. Putting all these together, we ought to gain a reasonable understanding of the physics and mathematics of ice crystals, and that's what snowflakes are made of. On the atomic level, a crystal is like a tiled wall, except that the tiles are configurations of atoms and they are arranged in three dimensions rather than two. So our starting point will be an easier problem with the same general characteristics as crystal physics— tiling patterns.

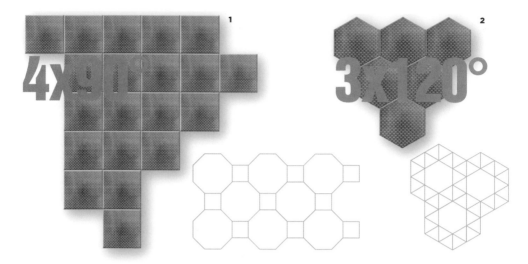

GEOMETRY OF TILINGS

One of the great constants of human culture is the tiled floor. The Egyptians laid stone tiles in regular patterns; the ancient Greeks and Romans often made similar regular patterns with their mosaics. Probably the simplest tiling pattern is a chessboard-like array of equal square tiles. Other than the square, the most regular tiling patterns use regular polygons—shapes with equal straight sides and equal angles at each corner. An equilateral triangle is a regular three-sided polygon; a hexagon is a regular six-sided polygon. Regular polygons can have any number of sides as long as it is bigger than three.

Which regular polygons can tile the flat surface of the plane, if we are permitted only one shape of tile? The answer is simple— equilateral triangles, squares, hexagons, and nothing else. The proof is equally simple. When you cover a wall or a floor with tiles, you use a small number of basic shapes (often just one, such as a square) but place copies of it in many different positions. The symmetry transformations that move a tile to another position and orientation are known as rigid

motions. Every rigid motion in the plane is a combination of a translation, a rotation, and a reflection, perhaps with some of these components being omitted.

Imagine a triangle and a rigid copy of it elsewhere in the plane. How can we move the first triangle to make it fit exactly on top of the second? First, we translate it until one vertex of the original coincides with the corresponding vertex of the copy. Second, (if necessary) we rotate the triangle about that vertex until one edge of the original coincides with and touches the corresponding edge of the copy.

There are now two possibilities. First: the original and the copy coincide. If so, do nothing—the task is complete without any final reflection. Second: the original and the copy are now mirror images of each other, reflected in their common edge. If so, reflect the original to make it coincide with the copy. Done!

In order to illustrate the mathematical classification of tiling patterns, I'll start with tilings by identical regular polygons. More than that: I'll insist that when they meet, their corners must coincide. No vertices of one lying on an edge of another. Given all this, we can see that

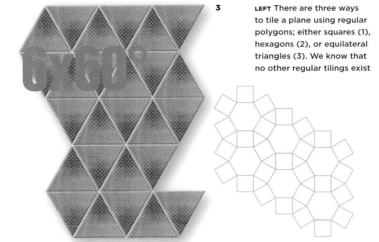

3

there are only three possibilities: equilateral triangles, arranged six to a vertex; squares, arranged four to a vertex—the standard grid pattern; hexagons, arranged three to a vertex, in a honeycomb.

Why are there no other possibilities? Why no pentagons, for instance? The key point is that the angles at each vertex have to fit exactly, with no gaps and no overlaps. So the corner of the polygon must be some whole number divisor of 360 degrees. This is so for equilateral triangles (60 degrees, one sixth of 360 degrees), for squares (90 degrees, one-quarter of 360 degrees), and hexagons (120 degrees, one-third of 360 degrees). But it doesn't work for pentagons (108 degrees): three pentagons leave a gap and four overlap. Nor does it work for any regular polygon with seven or more sides. So these three possibilities, called regular tesselations, are the lot.

If several different polygons are permitted, then the range of patterns increases considerably. Applying the same considerations about how angles fit together, there are precisely nine distinct so-called semiregular tilings, meaning that all tiles are regular polygons, they meet vertex to vertex, and the arrangement at every vertex is the same. Seven of these tilings are the same as their mirror images; two others come as a mirror-image pair. For example if we try to fit octagons together like bathroom tiles we find they leave square holes, which can be filled by square tiles. Similarly, hexagons meeting tip to tip leave triangular holes. Hexagons can be surrounded by alternating squares and triangles; dodecagons by triangles.

In artistic terms, tiling patterns reached their zenith in the work of Islamic artists, in decorations on mosques and other architectural works. Some of the most complex patterns make use of irregular polygons in a tiling. For example there is a pattern based on what seem to be regular polygons with 4, 5, 6, 7, and 8 sides. However, if these tiles really were regular, then the angles where the polygons met could not add up to 360 degrees, as they must if the tiles are to fit exactly. The artist has cunningly distorted the five- and seven-sided tiles to accommodate this discrepancy, but the changes are so slight that the eye does not notice the irregularities.

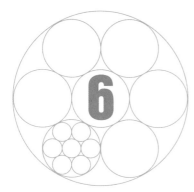

HEXAGON MAGIC

Because pentagons and any regular polygons with seven or more sides are forbidden in regular tilings, the main numbers associated with tiling patterns are 2, 3, 4, and 6. The number 6 was particularly important to the Pythagoreans. It is a triangular number: 6 = 1+2+3 (*see pp. 30–31*). It is also a "perfect" number—the (smaller) numbers that divide it exactly, without remainder, are 1, 2, and 3; they add up to 6, the number we started with. The next perfect number is 28, with divisors 1, 2, 4, 7, 14 that add up to 28. This is also a triangular number, equal to 1+2+3+4+5+6+7.

None of this is considered mathematically significant today. It's no more than a minor historical curiosity. However, the number 6 does have genuine mathematical significance, which is important for our snowflake story—it is the "kissing number" in two dimensions. That is, if you draw a circle in the plane and try to arrange several other same-sized circles so that each of them touches ("kisses") the first but none of them overlap, then you can fit exactly six circles around the first one. Try it with coins.

In three dimensions, where circles are replaced by spheres, the kissing number is 12. Try it with table-tennis balls—holding them in place is tricky, but double-sided tape is up to the task. At any rate, 12 balls will fit around a central one, but 13 won't, no matter how hard you try. In three dimensions, though, the fit is no longer exact; there are gaps between the 12 spheres, leaving room to slide them around.

In 2003 Oleg Musin proved that the kissing number in four-dimensional space is 24. Mathematicians know the kissing number for only two other spatial dimensions: 8 and 24. More sensible dimensions such as 5 and 6 remain an enigma, but in both of these cases we have a complete answer…which is insane.

ABOVE Identical circles in the plane, such as coins, pack together in a honeycomb—the tightest way to pack them. In this layout each coin "kisses" six others. If spheres are packed in space, the largest number of these that can kiss a given sphere is 12.

The story seems to rely on some rather bizarre coincidences; nonetheless it is proven beyond doubt that in 8 dimensions the kissing number is 240, and in 24 dimensions the kissing number is 196,560. I'm not going to explain what I mean by higher-dimensional spaces and spheres here—just relish the weirdness of it all. I love this kind of thing.

Back to two dimensions. Because the six kissing circles fit exactly, you can add more circles in exactly the same pattern to kiss them, and so on. The result is a honeycomb pattern of equal circles in which every circle is surrounded by six others. This pattern has even more symmetry than the sixfold snowflake. It has sixfold rotational symmetry about the center of any circle, not just about one point. It has reflectional symmetry about any line passing through the centers of two adjacent circles, or about any line that is tangent to two adjacent circles at the point where they touch. Not only that—it has translational symmetries. Choose any two circles and slide the whole pattern so that the first circle moves to the position of the second circle. The whole pattern fits exactly on top of its original position.

We've seen how Kepler convinced himself, in his *Six-Cornered Snowflake*, that the snowflake's characteristic symmetry occurs because the flake is made from identical tiny units arranged in a honeycomb pattern. He was almost right. There are identical tiny units—atoms of hydrogen and oxygen, the two constituent elements of water—and they are arranged in a lattice, but it isn't a honeycomb, not quite. (Tell you what it is later, OK? *See pp. 114–115.*) For pure thought in 1611, though, Kepler came amazingly close.

In nature we see a honeycomb pattern of basalt prisms in the Giant's Causeway in Northern Ireland. Basalt is solidified volcanic rock, and as the rock cools it shrinks and splits along flat planes. Because of the action of gravity,

in the vertical direction, a typical pattern is a honeycomb of vertical prisms. Of course, the honeycomb of the Giant's Causeway is imperfect, but no less impressive for that. Another place where nature uses hexagonal tilings is on the scales of snakes, lizards, and fish. A scale, after all, is a tile; and nature's economy forbids making scales with too many different shapes.

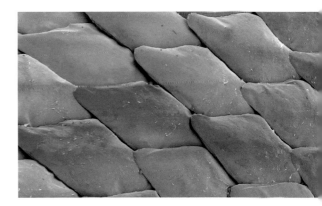

TOP AND ABOVE Very similar close-packed patterns occur in the scales of snakes and lizards (top), because it is an efficient way to cover a plane. A particularly dramatic example is the Giant's Causeway (above), where tall prisms of basalt rock are packed together in a honeycomb arrangement. These prisms formed as the molten rock cooled, and their close-packed arrangement is a consequence of that cooling process.

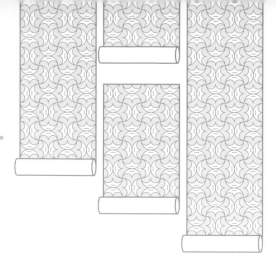

SYMMETRY & ART

Many art forms apply symmetry—sometimes in the precise mathematical sense, sometimes approximately. Symmetry is especially common in pottery and textiles; also wallpaper. In fact, wallpaper is one of the classic conjunctions of technology, mathematics, and art. The symmetries of wallpaper are closely related to tilings. Moreover, wallpaper is the mathematical key to crystal structure, as we'll shortly see (*pp. 114–115*).

In principle, the design on a roll of wallpaper could be anything. Hang up the paper temporarily, paint a design, wait for the paint to dry, take the paper down and roll it up. However, we live in a modern mass-produced world, so most wallpaper is produced repetitively by a machine that prints the design on rolls of paper using a rotating drum. This process naturally imposes constraints on the design, the most obvious one being that the same pattern has to repeat over and over again along the roll. However, a second kind of repetition is also necessary because successive lengths of paper must fit together neatly where they join. This condition means that the pattern must also repeat in a second direction—sideways. As every interior decorator knows, this sideways direction of repetition need not be horizontal—there can be a "drop" of the pattern from one length to the next. Nonetheless, the pattern must repeat in two independent directions, implying that it forms what mathematicians call a lattice. The regular and the semiregular tilings all have lattice symmetry. In wallpaper, the form of the pattern does not so much follow function as it follows the constraints of manufacturing technology.

So, regular tilings are lattices; and so, we will shortly see, are crystals. A general understanding of the theory of lattices and their symmetries looks as if it would be useful.

Such a theory, in two dimensions, was obtained by the American mathematician and educator George Pólya in 1924. In a joint work with Godfrey Harold Hardy and John Edensor Littlewood, he proved that there are precisely 17 distinct symmetry types for wallpaper—for two-dimensional lattices. These types subdivide into seven frieze patterns, now repeated indefinitely in parallel rows; two based on a rhombic lattice; three based on squares; and five based on a honeycomb. Careful choice of motif permits the inclusion or exclusion of certain mirror symmetries—the whole story is quite subtle. Remarkably, the analogous classification in three dimensions was found much earlier, around 1890, by three crystallographers named Evgraf Fedorov, Arthur Schoenflies, and William Barlow. Fedorov also did much of the two-dimensional case, but his work seems to have been forgotten by mathematicians.

All 17 wallpaper patterns were known to Islamic artists, where there has long been a tradition of using abstract designs, especially in decorating mosques and other significant buildings. Islamic artists generally saw strong links between the geometry of patterns and the geometry of the universe, so they used mathematical patterns to glorify the Creator. Probably they obtained their patterns by a mixture of intuition, experiment, and conscious analysis; at any rate, they seem not to have looked for a rigorous logical classification.

Indeed, there is no reason why they should have thought about their patterns in the manner of modern Western mathematicians, and plenty of evidence that they did not, such as the slightly irregular and strikingly beautiful "impossible tilings" already mentioned (*see pp. 74–75*).

The mathematics of symmetry entered Western art in a rather different way, during the Renaissance, when artists and mathematicians joined forces to work out the theory of perspective. Alberti's *On Perspective* is pretty comprehensive, though later artists such as Albrecht Dürer went on to build on his theories. Perspective involves the mathematics of symmetry because it relates rigid motions in three dimensions to the unavoidable distortions of a two-dimensional projection onto the artist's canvas. But it does not make central use of symmetric patterns as such, just the generalities behind them.

One artist who does make deliberate use of symmetric patterns is Maurits Escher. In 1922 Escher visited Spain and sketched in his notebook various designs from the Alhambra. From then on his work became ever more visibly influenced by abstract patterns and symmetries. His preliminary sketches for *Weightlifters* show that he based it on one of the Alhambra designs. He went on to invent new patterns of his own, sometimes in conjunction with mathematicians. His most characteristic contribution is the use of stylized animals as tiles.

TOP AND ABOVE The greatest exponents of lattice patterns in art and architecture are the Islamic artists, as can be seen in their tiled floors and carved screens (top). The artist Maurits Escher made extensive use of mathematical patterns, as can be seen in his *Angels and Devils* (above).

PACKING PROBLEMS

Modern mathematics has a habit of revisiting old problems and rethinking not just their answers, but what the original questions were. Kepler knew that when you pack equal-sized circles together in the flat surface of the plane, as tightly as possible, they form a honeycomb pattern. Or did he?

What Kepler knew was that you can fit circles together in a honeycomb, and when you do, there is no spare room left to move any of them about. That's one meaning of "tightly packed." Kepler would certainly have been able to prove these statements. However, there is a stronger meaning for "as tightly as possible"—that the gaps between the sheres are as small as possible. It begs the question, is the hexagonal arrangement the one that leaves the smallest gaps? And that question is a very different story. Only in 1910 did the mathematician Axel Thue offer a convincing proof that in this stronger sense there is no way to improve on the hexagonal packing. Even then, the proof slid over some tricky points, and the proof wasn't totally satisfactory until Fejes Tóth, also a mathematician, found a different approach in 1940.

Circle-packing still holds many mysteries, especially if you want to pack them in a finite container. For example what is the most efficient way to pack a given number of circles in a square box? The answers are known for any number of circles up to 20, but beyond that point our knowledge becomes hazy. There are some good guesses, but few proofs that they can't be improved. We do know that for 25 circles the best packing is the obvious 5x5 square array, and the same goes for 36 circles. However, there are better ways to pack 49 circles into a square than the 7x7 array. This should not be too surprising. After all, we know that in the infinite plane the hexagonal packing is better than the square one, so the way to get circles into a really big box is presumably to fill a large part of the box with a honeycomb pattern and then fiddle about near the edges.

The honeycomb has another important property. Suppose you want to divide the plane into lots of equal-sized regions, while keeping the total perimeter of those regions as small as possible relative to their area. That is, you want to create a two-dimensional foam that gives as much foam as possible for the least soap. It has long been assumed that the bubbles should be equal hexagons and the soap should form a honeycomb pattern. (I mean "long"—in the first century BCE the Roman, Marius Terentius Varro speculated that the honeycomb arises because it is the most economical design in terms of the quantity of beeswax needed.)

If you know that all the regions must have the same shape, as well as the same size, then you can prove that the honeycomb is the right answer. However, it's not obvious that all the regions should have the same shape, and mathematicians had run into difficulties with a similar question in three dimensions. What is the most efficient three-dimensional foam with equal-volume regions? Are all regions the same shape? The physicist William Thompson, who later became Lord Kelvin, asked the same question in 1887. He wasn't trying to understand foams, though. Instead, he was thinking about the structure of the "ether," the elusive medium that filled all of space and transmitted waves of light. Kelvin suggested that the regions should all have the shape of a tetrakaidecahedron, a polyhedron with 14 faces.

Kelvin's conjecture survived unchallenged until 1994, when mathematicians Denis Wheare and Robert Phelan found that foams made with two different shapes could knock 0.3 percent off the total area. We still don't know whether their

LEFT AND BELOW Identical cubes can be packed together to fill the whole space, and so can more complex shapes. However you pack spheres together, though, you have to leave gaps (below). All greengrocers know how to stack oranges efficiently (left), so it comes as something of a shock to discover that mathematicians spent more than 350 years wondering whether the greengrocers were right. The difficulty lies in proving that no alternative arrangement can pack the oranges more closely. At the end of the 20th century the greengrocers' intuition was finally vindicated, but only with the aid of computers.

proposal is the most efficient one, but we do know that their method—a massive computer search of possibilities—is unlikely to find anything better. A side effect of their counter-intuitive discovery was to make mathematicians start worrying about the two-dimensional case all over again. Must the regions of an optimal foam really have the same shape? In 1999 the American mathematician Thomas Hales showed that this time there is no need to worry. For once, intuition is right—the answer is a resounding yes: a honeycomb of identical hexagons. And in a brilliant tribute to the power of the human mind, his proof made no use whatsoever of a computer.

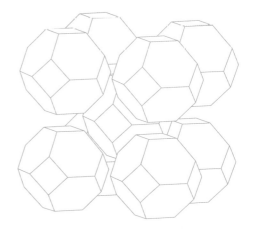

The ancient Greeks knew, and Euclid proved, that there are five regular solids. From left to right, these are the tetrahedron, cube, octahedron, dodecahedron, and icosahedron. These solids are highly symmetric. For example a cube can be rotated through multiples of 90 degrees, about any of its three axes, so that it occupies the same region of space that it did to start with. Crystals can assume the shapes of some regular solids, but not others. Threefold, fourfold, and sixfold symmetries are common in crystals. However, shapes with fivefold symmetry, such as the dodecahedron, are forbidden for a crystal lattice.

FIVEFOLD FORBIDDEN

In crystallography, 6 is a magic number. It is the highest degree of rotational symmetry possible for a lattice in two- or three-dimensional space. Two-dimensional objects can have sevenfold symmetry, and higher rotational symmetries—the simplest examples are the regular polygons with 7, 8, 9…sides. However, objects with this kind of symmetry cannot be crystal lattices. More interestingly, no lattice in two- or three-dimensional space can have fivefold rotational symmetry, either. The only degrees of rotational symmetry in such lattices are 2, 3, 4, and 6. This is the famous crystallographic restriction, first made explicit by amateur mineralogist

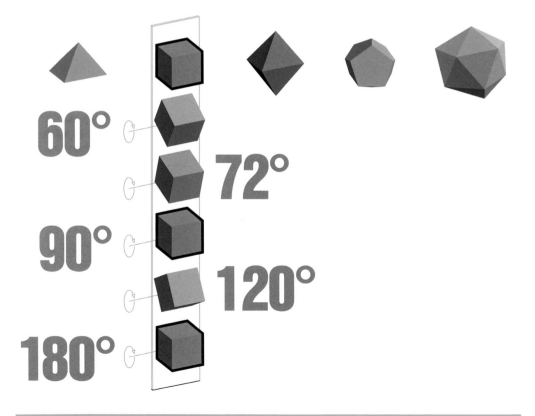

René Just Haüy in the early part of the 19th century. So 5 turns out to be a magic number for crystallography too, but in this case it is black magic, the forbidden art.

The ancient Greek philosopher Plato knew, from his studies of the works of Euclid and earlier Greek geometers, that there are precisely five regular solids—tetrahedron, cube, octahedron, dodecahedron, and icosahedron. These are shapes bounded by flat faces, such that every face is a regular polygon, all faces are identical, and the arrangement of faces at all corners is identical. A cube, for instance, has six square faces, all alike, and three of them meet at every corner, all at right angles to each other.

The "order" of a rotational symmetry is how many times you have to perform it to get back to where you started. Order 2 is a 180 degree rotation; order 3 is a 120 degree rotation; order 4 is 90 degrees, order 5 is 72 degrees; and order 6 is 60 degrees. For example the tetrahedron has rotational symmetries of order 2 and 3, the cube and octahedron have rotational symmetries of order 2, 3, and 4. The dodecahedron and icosahedron have rotational symmetries of order 2, 3, and 5. So there do exist highly symmetric three-dimensional objects with fivefold rotational symmetry.

Crystals can have the shape of a cube (common salt), an octahedron, and a tetrahedron. However, no crystal can take the form of a dodecahedron or an icosahedron because the crystallographic restriction rules out fivefold symmetry. This is a pity, for these solids are particularly beautiful, precisely because of that intriguing and forbidden fivefold symmetry.

Plato was heavily into number mysticism and he associated the tetrahedron with fire, the cube with earth, the octahedron with air, and the icosahedron with water—the four elements of the ancients. What of the missing dodecahedron? Plato associated that with the entire universe.

It is easy to see why fivefold symmetry cannot occur in a two- or three-dimensional lattice. As proof, suppose the contrary for a moment, that there is a crystal lattice with fivefold symmetry. One consequence of lattice geometry is the existence of a minimal distance between any two lattice points. Some two distinct points are separated by a distance that is less than or equal to the distance between any other two distinct points of the lattice. Now, this supposed fivefold symmetry of our pretend lattice means that each of these points is surrounded by five images of the other—the other point itself, and its rotations through multiples of 72 degrees. Elementary geometry then shows that there are two distinct points, one in each set of five, that are closer together than the original two points were. However, we know this to be impossible, since we supposed our original two points to already be at the minimum distance.

Mathematicians know that when an assumption leads to a logical contradiction, then that assumption must be at fault. Otherwise mathematics itself would contain logical contradictions and its treasured proofs would fall to pieces. Euclid called this method *reductio ad absurdum*—reduction to the absurd. Godfrey Harold Hardy, a leading mathematician in the early 20th century, compared the technique to an opening gambit in chess, where one player offers a piece to his opponent in the hope of gaining a long-term strategic advantage: "It is a far finer gambit than any chess gambit: a chess player may offer the sacrifice of a pawn or even a piece, but a mathematician offers the game."

The result of this technique is that the original assumption of the existence of a lattice with fivefold symmetry must be wrong. And that, plainly, means that there cannot be a lattice with fivefold symmetry. Thus do mathematicians prove some things are impossible.

OR MAYBE NOT

The "Mathematical Games" column in the *Scientific American* journal for January 1977, written by Martin Gardner, was a classic even by its author's high standards. It displayed to the outside world a remarkable set of tiling patterns invented by the Oxford physicist Roger Penrose. Penrose patterns bend the rules of crystallography close to breaking point, for they are based on pentagons. At first, they were fascinating material for pure mathematicians, but their practical application was nil. All that changed in 1984, when it turned out that nature had been making use of tilings like Penrose's, but in three dimensions. The result was the discovery of a new state of solid matter, the quasicrystal.

Penrose came up with two shapes of tile, known as kites and darts, obtained by dissecting pentagons into simple pieces. A kite is one-fifth of a pentagon, and a dart is what you have to add to a kite to make a rhombus. A set of kites and darts can be assembled into infinitely many different tilings of the plane. One such tiling, the Sun Pattern, has perfect fivefold symmetry—at its center sit five kites, recreating their parent pentagon. There is another fivefold symmetric pattern, too, the Star Pattern, with five darts at the center.

The laws of crystallography remain unbroken, however, because Penrose patterns are not lattices. But they come a lot closer than anybody had expected before Penrose's epic discovery. Penrose's patterns are not periodic, but they are quasiperiodic, meaning that any finite piece of them is repeated infinitely often—but not with perfect regularity. Although there are infinitely many possible tilings of the plane with kites and darts, they are all locally isomorphic, meaning that any finite subset that occurs in one tiling occurs somewhere or other, indeed infinitely often, in them all. If you were placed on a Penrose pattern and could explore only a finite region, you would be unable to tell which pattern it was. Another elegant property involves the golden number 1.618034, the limit of ratios of consecutive Fibonacci numbers such as 34/21 or 55/34. In a finite region of a Penrose tiling, the ratio of the numbers of kites to darts tends to the golden number as the size of that region tends to infinity. Overnight, the world was awash with Penrose-like, non-periodic tilings. Some were ingenious reformulations of Penrose's original system: Penrose himself was aware that the tiles could be changed to two different rhombi, with special "matching rules" for which edges could be fitted to which. Even tilings based on eightfold or twelvefold symmetry appeared. And to cap it all off, tilings of three-dimensional space with icosahedral symmetry were then discovered. Which just goes to show that whenever a new mathematical idea turns up, mathematicians worldwide come to grips with it, strip it down to its essential features, and try to find as many variations on it as possible.

Then, in 1984, the crystallographer Daniel Schechtman and colleagues announced that they had used X-ray crystallographic techniques to analyze a particular alloy of aluminum and manganese. This method is normally used to work out the atomic structure of crystals. It produces a diffraction pattern whose symmetry tells you which lattice the crystal is based on. But this particular alloy had a diffraction pattern with icosahedral symmetry. Because an icosahedron has fivefold rotational symmetries, the crystallographic restriction tells us that whatever the atomic structure of this alloy may be, it isn't a conventional lattice. It turned out that the atoms were arranged in a three-dimensional version of a Penrose pattern, an arrangement that was not periodic, but quasiperiodic.

This was a new form of matter, quickly named a quasicrystal. Later workers soon found other examples, for instance in an alloy of aluminum, lithium, and copper with six atoms of aluminum and three of lithium for every atom of copper.

The moral of the tale is: mathematical models have limitations. Behavior permitted depends on restrictions placed on the model. Nature works with the real thing, not with models, and does not always obey the restrictions mathematicians find convenient. The interesting thing is that we only realized that nature had transcended the limitations on crystal structure after the pure mathematics of Penrose patterns had been pored over and understood. Mathematics takes much of its inspiration from nature, but human imagination has an important role to play as well. Not for nature—the alloys would have made quasicrystals whether or not Penrose had invented his tiles—but for us.

RIGHT The physicist Roger Penrose discovered that if the requirement of strict lattice symmetry is relaxed, "crystalline" patterns with fivefold symmetry can occur. For example start with a rhombus (1) with angles of 72 degrees and 108 degrees and split it into two shapes, a kite (2) and a dart (3). Now it is possible to tile the plane with kites and darts in an infinite variety of ways. Two of these patterns have perfect fivefold symmetry: the sun pattern (4) and the star pattern (5). But all of the patterns have at least some elements of fivefold symmetry. Many variants of these "quasilattice" patterns have since been found, including some in three dimensions. A new state of matter, the quasicrystal, has also been discovered, in which atoms are always arranged in quasilattice patterns.

1

2

3

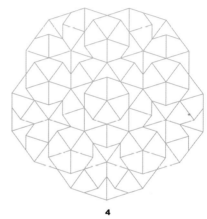

4

5

8

SPOTS & STRIPES

Abstractly, some animals walk around covered in wallpaper. Hairy wallpaper. Their markings have lattice symmetry, give or take distortions caused by the odd stomach or leg. The Scottish zoologist D'Arcy Thompson, in his celebrated work *On Growth and Form*, writes: "The zebra is striped that it may graze unnoticed on the plain, the tiger that it may lurk undiscovered in the jungle; the banded Caetodont and Pomacentrid fishes are further bedizened to the hues of the coral-reefs in which they dwell. The tawny lion is yellow as the desert sand; but the leopard wears its dappled hide to blend, as it crouches on the branch, with the sun-flecks peeping through the leaves."

We are told that the leopard cannot change its spots, but how did it get its spots to begin with? How did the zebra get its stripes? What are the patterns for, anyway?

Some reasons are probably evolutionary. Distinctive markings let you distinguish your own species from a different one. This is useful when seeking a mate and crucial when avoiding predators. It seems clear that some of the more lavish markings on birds—the plumes of the peacock, the elaborate and colorful curlicues and quills of birds of paradise—evolved through sexual selection. This is a process in which slight random preferences by females for particular features in males become amplified in a positive feedback loop, driving the males to ever-greater excesses of plumage. Indeed, the more difficult it is for the male to stay alive with all this excess baggage, the more likely it is that female preferences will tend in that direction—because a male that can function with such an enormous handicap must have exceedingly "good genes." Evolution takes place over such long periods of time, however, that most stories of this kind are virtually impossible to justify scientifically, so at best we should consider them as guidelines.

RIGHT There are lattice symmetries in the animal kingdom, although these are not as regular as those found in crystals. Stripes (in an idealized mathematical representation) correspond to the simplest lattice pattern in the plane—alternating bands of different colors. Stripes are found throughout the natural world—in tigers, fish, butterflies, zebras. Mathematics helps us understand how stripes and other animal markings form, and why stripes are such a common pattern in living creatures.

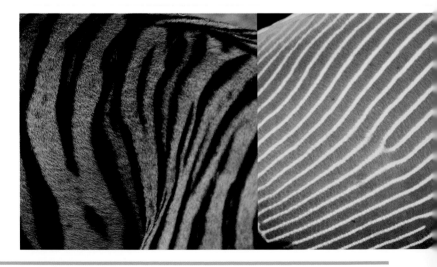

Stripes are bold and dramatic—easy to recognize. Many animals, such as zebras, tigers, wild piglets, and raccoons, have stripes, but there are stripes elsewhere in the animal kingdom, too, notably on seashells. The maximum keyhole limpet is a striped cone with a hole in the top, like a circus tent, with brown and white stripes radiating from the cone's tip. The striped bonnet, found in the Indian and Pacific oceans, has stripes that run at right angles to its spiral coils, the closely related smooth scotch bonnet, found in the Caribbean, has stripes that run parallel to its coils. These are the two typical directions for stripes on seashells. The gyrate frog shell has what looks like stripes, but is in fact a single brown stripe and a single white one coiled around and around, following the shape of the shell. Eliot's volute, found in the sands of South Australia, has widely spaced thin brown stripes on a pale background. The genuanus cone from West Africa has broken black stripes, which look like dotted lines.

The greatest exponents of the stripe as a decorative add-on are tropical fish. And there seems to be no end to what they can do with them. The blue-striped grunt has wonderful wiggly blue stripes running the length of its body. They are edged with black, on a yellow background. The French angelfish has half a dozen thin yellow stripes running vertically on a black background. The sergeant major is a silver fish with five black bars across its body. The grumpy-looking Nassau grouper is a light gray fish with dark gray stripes, which run around its body but along its angular head. The reef squirrelfish is mainly red with thin white stripes running lengthwise. And the sand perch seems unable to make up its mind, for it has stripes in two directions—thin blue ones running lengthwise, overlaying black and white stripes at right angles.

What is all this with stripes? One possibility is that stripes are in a sense the simplest possible pattern in the plane. Really, a set of stripes is no more than a one-dimensional pattern—alternating intervals of two different colors—that has been smeared sideways to cover the entire plane. If so, mathematicians could study stripes by reducing everything to a one-dimensional cross section.

They do.

SPOTS

Next in order of geometric complexity, after stripes, come spots. Seashells often have spots, for example the sieve cowrie; so do tropical fish, such as the spotted trunkfish or the yellowtail damselfish. Animals that have stripes often have close relatives with spots, as the big cats exemplify. The leopard is notoriously spotted, along with the cheetah, jaguar, margay, and snow leopard. On closer examination, their spots often turn out to be complicated structures in their own right. It also looks as though spots bear some relation to stripes—they often seem to be stripes that have failed to join up properly.

The cheetah's body is covered in sandy-colored fur, liberally dotted with roughly circular black spots. The spots are of slightly different sizes and their arrangement is not particularly regular; neither, though, is it random. They are spread fairly uniformly over the body, separated from each other by roughly equal distances. Random arrangements of dots would have clumps of dots in some places and big gaps elsewhere.

What strikes the eye, looking at a cheetah, is a tendency for the spots to line up in rows. This might perhaps be an optical illusion, but a glance at the animal's tail suggests there's more going on than that. At the top of the tail there are rings of spots that run right around. The spots in each ring are very close together, while the rings themselves are clearly separated. Moving back along the tail toward its end, we find that the spots in each ring start to overlap, then merge, and finally they join together into a seamless ring—a circular stripe. Roughly half of the tail is decorated with parallel ringed stripes.

To anyone familiar with mathematical pattern formation, an irresistible conjecture automatically springs to mind. Lines of spots are what you would expect in a system that was "trying" to form stripes, but where the stripes themselves are unstable. Let me explain with a seagoing analogy. We all see lines of waves rolling up a beach and breaking on the shore. Metaphorically speaking, these lines are moving stripes of water. If we colored the sea red along the tops of the waves and blue along the troughs, then we'd see parallel red and blue stripes.

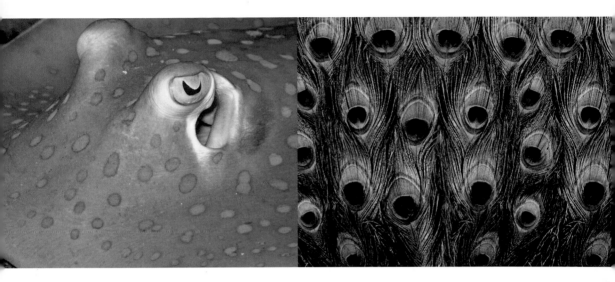

Mathematics thrives on metaphors, and it does so by turning them into more significant similarities. What we have with the wave analogy is a similarity of processes, and it leads to the same general class of patterns because these are the typical patterns in all such processes. The process is the formation of waves on a uniform substrate. In the sea, the substrate is the flat undisturbed surface of the water, and the process is driven by currents and winds. On a zebra, the substrate is the distribution of pigments in its hair, and the process is driven by chemistry. In one case, we see the wave because of its shape; in the other, we see it because of its color. Mathematically, there is no essential difference.

STRIPES

The stability of a pattern depends on its reaction to disturbances. Put simply if it retains its form when disturbed, it is stable; if not, it is unstable. One standard way for a linear wave—a stripe—to become unstable is to develop a ripple that runs along it. The previously parallel edges of the stripe become wavy, so that it alternates between being wide and being narrow; then the narrow parts of the wave come to bits completely and the wave then breaks up into a long line of clumps. This is exactly what the cheetah's spots seem to be doing near its tail, and possibly elsewhere on its body, too. The instability in the stripes on a cheetah, though, is chemical rather than fluid-dynamical.

The spots of the leopard are also arranged in long lines, so the same general pattern-forming mechanism is probably at work. Our ignorance of the precise details (what kind of chemistry is at work here?) is driven home by the remarkable form of each individual spot. The leopard's background fur is light and sandy-colored. Each of its spots has a brown center, which is surrounded by what, at first sight, looks like a black ring. But the ring is clumpy and often breaks up into three or four black spots. The snow leopard, on the other hand, has a very different coloration, gray and cream, but like its cousin, its spots are equally intricate.

Any mathematical theory of animal markings must take into account and explain all of this.

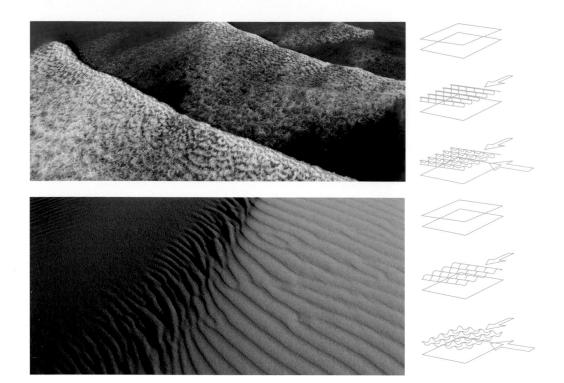

WAVES

The relation between stripes and spots suggests that we might make progress toward that elusive snowflake by looking at the mathematics of waves. For animal markings, these are presumably waves of chemicals, but the great thing about waves is that the same general mathematical ideas apply to all. This allows us to exploit one of the great features of mathematics—its suitability for technology transfer.

We can look at one kind of wave, preferably an easy idealized one to experiment on and think about, and from that deduce principles that apply to lots of other waves, including ones that are physically far less accessible. That may sound like an extreme statement. Every mathematician knows that the equations of fluid flow differ considerably from the equations of diffusing chemicals. Nonetheless, there are some deep commonalities and the most robust of these is the effect of symmetry on pattern formation.

Let me give you an example involving two very different physical systems—oceans and sand dunes. The commonest pattern in a large flat square of fluid (a mathematician's idealized ocean) is a set of parallel waves. The commonest pattern in a large flat square of sand (a mathematician's idealized desert) is a set of parallel dunes. The next commonest pattern in a large flat square of fluid is a grid of "spotty" peaks and troughs. The next commonest pattern in a desert is a grid of "spotty" barchan dunes. As you can see these similarities go beyond metaphor. Both systems, when idealized, have the same symmetries, and this implies that they select their patterns from the same mathematical catalog.

The analogy is remarkably close. With the right laboratory apparatus, it turns out to be even closer. At the end of the 19th century the French scientist Maurice Couette introduced a fascinating experiment to the scientific community in which a fluid is confined between two cylinders, the inner one being made to

LEFT Sand dunes in the desert
reduce, mathematically, to
patterns of waves in a plane.

rotate. Later the British applied mathematician Geoffrey Ingram Taylor took the idea further, so the experiment is now called the Couette-Taylor system. We can think of a cylinder as a rolled up square, and as long as we're interested in periodic patterns this mathematical reinterpretation makes little difference to the catalog.

The only pattern that Couette was interested in was a very dull one—no pattern at all. This one occurs when the cylinder is rotated slowly. At higher speeds, as Taylor predicted, we find a stack of ring-shaped vortices, today called Taylor vortices. At higher speeds still, the vortices become wavy (in effect, they start to break up into spots). Later experimentalists made the outer cylinder rotate as well and found that if it rotates in the opposite direction to the inner one, the result is rotating helical waves like the stripes on a barber's pole. There are other fascinating patterns, too—wavy vortices, twisted vortices, interpenetrating helices, even spiral turbulence—but for our purposes these two will suffice.

We can transfer this mathematical technology to sand dunes. We conceptually slice the cylinder vertically and unwrap it into a flat sheet, taking the patterns with it. What do we get? Couette's dull patternless fluid becomes dull patternless desert—flat desert without any dunes. Taylor's stacked vortices become stripes—parallel rows of heaped sand, known as transverse dunes. Wavy vortices become barchan dunes, stripes that

have started to break up. And helices become stripes placed at an angle, known as linear dunes, or seifs.

There are lots of technicalities to take care of before this analogy can be turned into solid science—not the least of which is the absence of any good solid equations for sand—but in spirit, what I've told you is much more than a visual pun. Similar analogies work for all sorts of pattern-forming systems in the plane—from liquid crystals to visual hallucinations, but with some extra mathematical considerations that are even more technical. The main point is that the symmetries of a system go a long way toward telling you what patterns it can or cannot form. More careful analysis, based on detailed physics, will then tell you which of those possible patterns will actually form, and under which circumstances. So we can get the best of both worlds—unity and diversity.

BELOW Different patterns of waves arise under different conditions (1): the pattern depends on the speed and direction of the wind. The same list of patterns arises in a very different realm of physics. In the Couette-Taylor experiment (2), fluid is sandwiched between two rotating cylinders. Depending on the rotation speeds, various patterns form. The underlying symmetries of this system are the same as those of sand dunes: to see this, it is necessary to open up the cylinder and roll it flat. It now becomes a plane, and the rotation has the same effect as wind blowing over the surface of the plane.

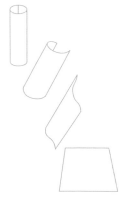

1

2

WEATHER PATTERNS

Patterns rather similar to those in the Couette-Taylor experiment (*see pp. 90–91*) also arise when a layer of fluid is heated. Hot fluid rises because it is less dense than colder fluid. So what happens if you start with a shallow layer of fluid and heat it uniformly from below? It can't all rise, or the fluid would just leap into the air. Instead, when the critical temperature is exceeded, the fluid breaks up into some regions where it flows upward, cooling down when it gets to the top, and other regions where the cool fluid falls back down again, to be warmed up and continue the cycle. The Bénard experiment, a laboratory version of this scenario, shows that the fluid can form striped patterns, checkerboards, honeycombs, and more.

This tendency of heated fluid to move is known as convection, and the regions that it breaks into are called convection cells. This is a major feature of weather systems, with the heat being supplied by the Sun.

Since our objective is the snowflake, it's worth taking a look at weather patterns and how they arise.

On small scales, we can pretend the Earth is flat. The atmosphere forms a relatively thin layer. The Earth rotates, causing winds as the Sun rises and sets and the atmosphere heats and cools. The Sun warms the air during the day but not during the night, when the heat is radiated back into space. The atmosphere is a mixture of gases and water vapor—these days with a liberal dash of pollution—and weather is what happens when the atmosphere obeys the laws of physics.

The laws seem to allow rather a lot of things.

The surface of the Earth is rough, and weather is affected by terrain. As winds pass over the top of a mountain range, they often form a series of waves on the lee side—the air moves up and down in sinusoidal curves. At the peaks of these curves clouds can form. The result is a series of cloudy stripes running parallel to the mountains, known as wave clouds.

Convection cells in the atmosphere are responsible for one of the commonest forms of cloud, known as cumulus. Warm air near the

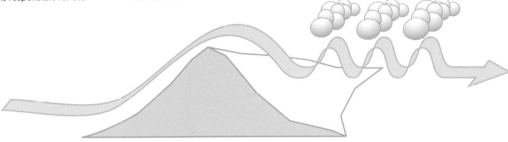

ground picks up moisture from vegetation, rivers, and lakes, and then rises. But the upper regions of the atmosphere are cold, and cold air cannot hold as much moisture as warm air, so some of the moisture condenses out, forming wooly white clouds. Once the air has cooled enough, it stops rising, an effect known as inversion. If inversion occurs at about 5,000ft (1,500m), as it typically does in summer, the cumulus cloud remains short and wide—this is the fair-weather cumulus. In the absence of such an inversion, the cloud can grow to twice that height, in which case its top rises up into the level at which ice crystals can form. This could provoke a shower, as the ice gets caught in the convective swirl and is carried down to warmer levels, where it melts before falling to earth. In more extreme conditions, the cloud can push higher still—5 miles (9km) in temperate zones, 9 miles (15km) in the tropics—developing into cumulonimbus, the classic thundercloud. The top spreads outward to form a dense mass of cirrus (ice) cloud which often flattens at the top to form an anvil shape because it cannot rise any higher

and the high winds spread it out. Water vapor condenses onto ice crystals at the top of the cloud, which grow to form solid lumps of ice. These can fall as hail or, if they melt on the way down, as rain. Hailstones are often layered, like an onion, perhaps as a result of making several trips through the cloud and gaining a new layer on each passage.

Water and ice are not the only things that circulate in a storm cloud. One of the most dramatic is electricity. The top of the cloud acquires a strong positive electric charge, while the lower regions are mostly negatively charged with the odd positive patch, often where the rain is heaviest. The electrical tension builds and builds until eventually something has to give way and a flash of lightning leaps from cloud to cloud or from cloud to ground. The sudden displacement of the air creates a shockwave— thunder. And sometimes, instead of rain, when the atmosphere is cold enough clouds produce snow, making it clear that more than one pattern-forming process, on more than one scale, goes into the formation of a snowflake.

TURING'S TIGER

Fine—so we see stripes and spots in the clouds, created by purely physical processes and governed by mathematical rules. But animal markings involve biological processes too. They are patterns of pigment, and pigments are proteins made by genes, so the chemical ingredients for the tiger's stripes and the leopard's spots must depend on genetics. Nonetheless, there is a case to be made that the patterns themselves, or at least the general range of available patterns and the principles underlying which ones will occur, are governed by mathematical rules that operate in conjunction with genes.

In 1956 the mathematical logician Alan Turing showed in a series of highly complex theories that systems of chemicals reacting together and diffusing through tissue can create spontaneous patterns. He called these chemicals morphogens—shape creators. When Turing first published his ideas, they were purely theoretical, but a good example of real-world Turing patterns soon came to the attention of chemists, the so-called Belousov-Zhabotinskii reaction. If certain chemicals are mixed together and placed in a shallow dish, they form a uniform brownish sludge. If you wait a few minutes, though, tiny blue spots appear, apparently at random. The blue spots spread and their centers turn red. Soon the dish is full of concentric rings of red and blue chemical, known as target patterns. Shake the dish a little, and they turn into slowly twirling spirals.

Not the patterns you find on most animals, mind you. But it turns out that Turing's equations can produce a huge range of different patterns, including stripes, spots, dappling, and much else—even complex spots like the leopard's. One problem with the patterns in the Belousov-Zhabotinskii reaction, though, is that they aren't static, they move. We don't see zebras with moving stripes, or leopards with moving spots. However, Turing showed that his equations can produce both stationary patterns and moving ones, depending on the reactions and the rates at which the chemicals diffuse.

Whatever the precise mechanism of biological pattern formation might be, it is not simply that the pigments on the animal's skin or in its fur react and diffuse. It has to be some kind of multi-stage process. Moreover, it takes place in the embryo, not in the adult animal. Even if the

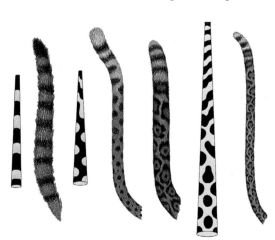

LEFT Alan Turing modeled animal markings using mathematical equations for the diffusion of chemicals. When his equations are applied to tapering cylinders the patterns that form are very similar to those on the tails of big cats—the leopard, jaguar, cheetah, and genet.

embryo does not exhibit clear patterns, they have to be present in some cryptic form. To Turing, the key point was that his system produced the right kinds of pattern. If pigments are deposited according to the peaks and troughs of parallel waves you get stripes, more complex systems of interfering waves produce spots, and so on.

Turing's early mathematical equations were a bit too far removed from real biology to provide accurate models. Modern genetics fills in a different piece of the puzzle. It explains the production of proteins, but it doesn't explain how the proteins are assembled to form an organism or—crucially—why nature so often prefers mathematical patterns. Recent research combines both points of view and shows that in many cases Turing's equations fit experiments better than rival theories that had been preferred by biologists.

To see the difference between the two approaches and how they both fall short of reality, imagine a vehicle (corresponding to a developing organism) driving through a landscape (representing all the possible forms that the organism might take, with valleys corresponding to common forms and peaks to highly unlikely ones). In models like Turing's, once you have set the vehicle rolling, it has to follow the contours of the landscape. In contrast, the current view of the role of DNA sees development as an arbitrary series of

instructions: "Turn left, then straight ahead for a hundred yards, then turn right…" Any destination is possible given the right set of instructions.

The true picture, however, must combine genetic switching instructions and free-running chemical dynamics. If a car driving through a landscape follows a series of instructions, it will drive into a lake or fall off the edge of a cliff. A car with a driver has more freedom in selecting destinations than a free-running vehicle without any controls. In the same manner, an organism cannot take up any form at all—its morphology is constrained by the laws of physics as well as by its DNA. But the DNA instructions can make arbitrary choices between several different lines of development that are all consistent with the physical laws. It is not DNA alone, or dynamics alone, that controls development. It is the interaction, like a landscape that changes shape according to the traffic that passes through it.

SLIME MOLD SPIRALS

The humble slime mold illustrates the nature of the problem rather nicely—it's not just your genes that matter, it's what you do with them. This tiny, unintelligent creature manages to create the most spectacular spiral patterns. To what extent are these patterns encoded in its genes? Is there a gene for spirals?

To answer this question, we need to know how the slime mold makes its spirals. And the first thing to realize is that the spirals are a collective activity. Slime mold is not one amoeba, but a colony. Its life cycle begins with a microscopic spore, a dried-out amoeba that can be blown on the winds until it finds somewhere nice and moist. It then turns back into a genuine amoeba, hunts for food, and starts to reproduce by splitting in two whenever it gets big enough. Soon there are lots of amoebas. When the crowd gets too big and food runs low, the amoebas separate into small patches. All the amoebas in

one patch crowd together; and as the crowd makes its way toward its common destination, they form elegant spirals—which, incidentally, rotate slowly.

As time passes, the crowd of amoebas becomes denser and the spirals get more and more tightly wound. Then they break up into streaming patterns that look like roots or branches. The streams thicken, and as more and more amoebas try to get to the same place they pile up in a heap, familiarly known as a slug. The slug isn't an organism—it's a colony. Nonetheless, it moves as if it were a single organism, looking for somewhere dry to "reproduce." When it finds a dry spot, it attaches itself firmly to the ground and puts up a long stalk. The rest of the amoebas form a round blob at the top of the stalk. The amoebas in the fruiting body turn into spores and blow away on the wind and the cycle starts all over again.

It may sound complicated, but this could be an illusion—Thomas Höfer and Martin Boerlijst,

LEFT AND BELOW Amoeba arithmetic is very unusual: they multiple by dividing (below). Slime mold is a colony of amoebas, which has two ways to reproduce. Slime mold amoebas divide and spread themselves out into a slimy film. But when the population becomes too large, slime mold will congregate in patches. Some of the amoebas dry out to form spores held in a round blob (left), and the rest make a stalk for it; then the spores blow away on the wind.

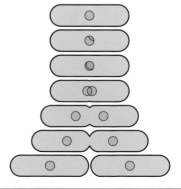

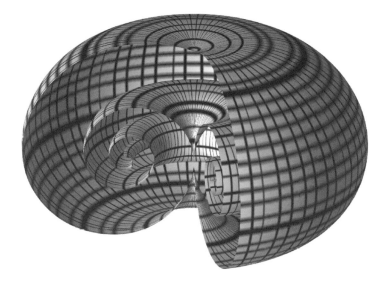

RIGHT Slime mold amoebas can also move around in a "slug" using three-dimensional scroll waves, which can also be observed in chemical reactions.

both mathematical biologists, have discovered a relatively simple system of mathematical equations that reproduces both the spirals and the streaming patterns. The main factors responsible for the patterns are the density of the amoeba population, the rate at which the amoebas produce a chemical known as cyclic AMP, and the sensitivity of individual amoebas to this chemical. Roughly speaking, each amoeba "shouts" its presence to its neighbors by sending out cyclic AMP. The amoebas then head in the direction from which the shouts are loudest. Everything else is a mathematical consequence of this process.

The mathematical biologist Cornelius Weijer has shown that very similar equations can also model the movement of the slug. This is a three-dimensional problem, and the answer involves a remarkable three-dimensional wave called a scroll wave. Art Winfree, another mathematical biologist, predicted that it is possible for such waves to occur in the Belousov-Zhabotinskii reaction in three dimensions, and experiments have since detected them. A scroll wave is similar to a spiral wave, but with an extra twist. Literally. Imagine taking a sheet of paper and rolling it up so that in cross section it forms a spiral. Next, imagine that it is very flexible paper, able to stretch and bend without

crumpling. Bend the two ends around until they meet—now you have a sort of doughnut shape with a spiral cross section, and that's a scroll wave.

This is almost the scroll wave, but not quite. To get that, you have to twist one end of the paper through a complete circle before joining it to its fellow. The ends will still match, since you've made a complete twist, but now the spiral cross sections turn through 360 degrees as you move around the doughnut.

The final ingredient is to recall that in two dimensions the Belousov-Zhabotinskii spirals are not static, they rotate. All those spiral cross sections rotating in step with each other—that's a scroll wave. Its curious, twisty-twirly rotation is just what's needed to make the slug lollop its way across the ground, looking for somewhere to root and put up its fruiting body.

Most of the slime mold's genes just tell it how to be an amoeba. The genes that help it to form patterns tell amoebas how to send out chemical signals, how to sense them, and how to respond to them—but the actual patterns themselves are not specified in the genes. Instead the patterns emerge from the mathematical rules that the chemical signals and the amoebas are obeying. The life cycle of the slime mold owes as much to mathematics as it does to genetics.

STRIPES THAT MOVE

Old theories never die…

For years it was thought that the big defect in Turing's theory was its tendency to produce moving patterns much more readily than stationary ones. Chemists looked hard for Turing patterns that didn't move, but couldn't find any. Moving patterns, like those in the Belousov-Zhabotinskii reaction, were relatively easy. Stationary ones—not a whisper.

Unfortunately, it's obvious that the patterns on living organisms don't move. Big problem.

But…

Sometimes the patterns on organisms do move. They move rather slowly, which is why we don't generally notice, but they do move. In fact, Turing's model correctly predicted that if they do move, they should move slowly.

The organism that caused this latest change in thinking is a small tropical marine fish, the angelfish *Pomacanthus*. The young are about ¾ in (2cm) long, adults three to four times that length. There are many different *Pomacanthus* species and they exhibit a variety of patterns. *Pomacanthus semicirculatus*, for example has curved stripes that run vertically down the body, while *Pomacanthus imperator*, the emperor angelfish, has horizontal stripes that run the length of its body. Over time, the stripes of both species change their pattern. This is especially clear for *Pomacanthus semicirculatus* because the young fish have only three stripes, but adults have 12 or more, so somehow the number of stripes has to increase as the fish develops. In fact, the changes actually occur in a rather curious manner.

Start with a juvenile fish which has the usual three stripes, and watch it grow. At first, the stripes expand with the fish, becoming more widely spaced—this is what you would expect if the pattern were laid down once and for all. But at this stage, relatively suddenly, new stripes begin to appear between the original ones, restoring the original size of spacing. At first they are thinner than the old stripes, but they gradually thicken. When the body length reaches about 3in (8cm), the process is repeated a second time.

This sequence of changes has been modeled using reaction-diffusion equations by the Japanese mathematical biologists Shigeru Kondo and Rihito Asai. Their model involves just two chemicals and assumes that the underlying tissue consists of a row of cells, some of which duplicate every so often. The results reveal a natural pattern of stripes, which widen without changing their number until the tissue becomes

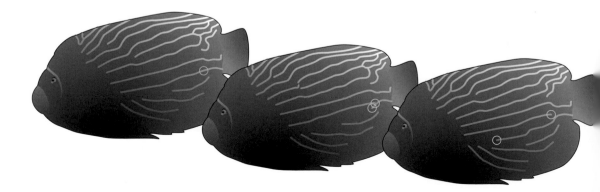

sufficiently large, at which point the number of waves doubles, with new stripes appearing between the old ones. This is exactly as Turing predicted fish.

An even more dramatic scenario arises in the horizontal striping of the emperor, *Pomacanthus imperator*. It also develops additional stripes as it grows, but some of the stripes "unzip" and split into two. This type of wave rearrangement is known to physicists by the term "dislocation" and it is widely observed in a variety of systems. In particular, it occurs in reaction-diffusion systems. To say that the stripe "unzips" is a slight simplification because it suggests that a single stripe turns into two by developing a Y-shaped branch point. It can happen this way, but there are also more complicated dislocations in which stripes rearrange themselves by disconnecting and reconnecting, and Kondo and Asai saw these too. They used the observed spacings of stripes in the fish on which to base an estimate of the diffusion rates for their hypothetical morphogens (shape creators), and the results are within the range you'd expect if each morphogen were some kind of protein molecule.

Changes occurring in stripe patterns in angelfish are thus consistent with mathematical equations of Turing's general type, and—crucially—they are not what you would expect

in patterns that are simply laid down, cell by cell, by arbitrary genetic switches. So it looks as if something mathematical, accessing the laws of physics and chemistry, must be going on in addition to genes switching each other on and off.

Whether it takes the physical form assumed by Turing—diffusing waves of chemicals—is not clear, though. One possibility is that Turing's equations are actually modeling some approximation to the progress of genetic switching through the organism. When a lot of people in a sports stadium stand up and wave their arms as soon as they see people on their immediate left doing the same, then it looks like a wave of movement but everyone ends up in their original seat. No physical substance actually travels through the crowd. So Turing patterns might be the consequence of genetic planning, not chemicals that react and diffuse. No matter—the catalog will be the same.

BELOW In some circumstances, Turing's equations predict moving patterns of stripes instead of static ones. Animals with stripes that move? It seems unlikely. However, observations by Japanese scientists have shown that the stripes of the emperor angelfish do in fact move—over periods of many weeks. Not only that: the patterns break apart and rejoin in exactly the manner that the equations predict.

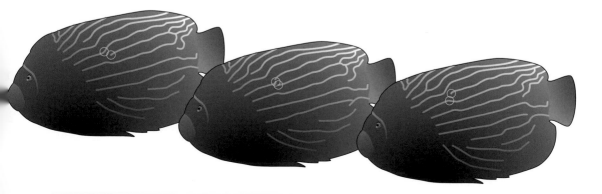

MATHEMATICS & BEAUTY

Let's pause and take stock.

As a child, I was lured by the snowflake's beauty; now I seek its resolution in mathematics. Is this wise?

It may seem surprising to combine the words "mathematics" and "beauty" in one phrase. Most people's mental image of mathematics is page upon page of complicated "sums"—not an especially beautiful sight. I sympathize, believe me. But that's arithmetic, not mathematics (I'm quite passionate about this). Those symbols on the page come no closer to the true beauty of the subject than the staves and sixteenth notes of musical notation come to a complete Beethoven symphony. The beauty of mathematics also lies not in its notation but in its ideas; not in its five-finger keyboard exercises but in its symphonies.

There seem to be two kinds of mathematical beauty—logical and visual. The philosopher and mathematician Bertrand Russell once described the beauty of mathematics as being "cold and austere," and here he was referring to its logical beauty. To someone who understands the ideas, a mathematical proof can resemble a symphony in logic. This kind of beauty is intellectual and not easily accessible to the uninitiated.

Visual beauty, in contrast, has an immediate and direct appeal even to a casual observer. This book is littered with examples of attractive shapes and patterns produced by mathematical processes. The beauty of the snowflake is mathematical—it appeals to our sense of symmetry and complexity, and these are the essence of mathematics.

The relation between mathematics and beauty is genuine, but elusive. There seems to be no prospect of inventing a Calculus of the Beautiful (not that this has dissuaded some brave and foolhardy souls from trying to find

one). Moreover, idealized mathematical patterns tend to be just a bit too regular when compared to the natural world or to art to be considered beautiful. Nonetheless, our visual senses seem to be attracted to intricate, repetitive patterns—symmetry. We surround ourselves with such images. Look at wallpaper, curtains, carpets, upholstery, pottery, even architecture.

What is it about symmetry that so appeals to our senses?

The human mind seems to enjoy repetition—up to a point. Children love being told the same story over and over again. Music, at its lowest level, consists of rhythmically repeating noises. At its highest, we find theme and variations, interwoven patterns of almost-repetitions. The brain evolved in a world where the ability to recognize patterns offered increased survival value. An understanding of the seasons allows people to find food throughout the year. Pattern recognition lets us distinguish snake from vine, wasp from butterfly.

Our minds seem to be built from a large number of intercommunicating modules, which evolved because they contributed to our survival. Our feeling for beauty and our ability to do mathematics seem to be side effects of the activity of these modules. Recently a survey carried out on the Internet asked what kinds of painting people of various nations preferred. With one exception—the Dutch—everyone liked a landscape with water, distant hills, animals, and a few trees (not too many). In England the animals were cows and in Kenya they were hippos, but the general attitudes were identical.

Digging deeper, it has since turned out that most people can recognize such a scene so rapidly that this ability must be a built-in reflex—a mental short-circuit like the one that closes our eyes if an object approaches them at speed. These reflexes evolved because rapid responses were better than accurate ones.

So what is the survival value of a landscape? Safety. Such landscapes possess all the ingredients required by early hominids. Food, water, shelter. You can climb a tree, but you don't want there to be so many trees that predators can easily hide behind them.

It's a nice theory. Right or wrong, it shows how curious our aesthetic sense must be; and it illustrates how this sense relates to preferences for certain patterns and the ability to detect them. Mathematics is a systematic, partly conscious technique which we have invented to exploit our highly evolved mind's eye for pattern. It is only reasonable to expect a strong link between mathematics and beauty.

ABOVE Millions of years ago landscapes with water, a few trees, some animals, and mountains in the distance had evolutionary survival value. Today we just enjoy the scenery.

9

THREE DIMENSIONS

Despite the rich variety of patterns we've encountered, we've hardly begun. So far, almost everything has been happening in the plane. The crystal structure of ice, one of the keys to the snowflake, is three-dimensional. And in three dimensions, there's room for a lot more to happen. The ingredients for three-dimensional symmetry are much the same as in two dimensions, so we can build on the intuition we've developed so far; but those ingredients can be combined in new ways.

Now is the time to return yet again to one of the high points of ancient Greek geometry—the classification of the regular solids. Only this time we will focus on their symmetries. As we've seen, the Greeks could prove that there are exactly five regular solids: the tetrahedron, formed from 4 equilateral triangles; the cube, formed from 6 squares; the octahedron, formed from 8 equilateral triangles; the dodecahedron, formed from 12 pentagons and the icosahedron, formed from 20 equilateral triangles. Not only that—those amazing Greek mathematicians could prove that no other regular solids exist. This fact forms the climax of Euclid's *Elements*.

To the Greeks, though, the classification was just a list of possible forms. Modern mathematics has recast it as a list of possible symmetry types, and its influence has been immeasurable.

To see what's involved here, think of the most familiar regular solid, the cube. How many symmetries does a cube possess, and what are they? A cube is a beefed-up square, so we can use a square for inspiration. A square has four rotational symmetries (through 0 degrees, 90 degrees, 180 degrees, and 270 degrees) and four reflectional symmetries (central axes and diagonals). We can apply all eight of these symmetries to a cube by choosing one face— a square, of course—and moving the whole cube to reproduce the symmetries of that square. For descriptive purposes only, imagine that square is red. In order to rotate the red face, we rotate the whole cube. In order the reflect the red face, we reflect the whole cube. So far, so good, but there are some crucial differences. In two dimensions, a rotation fixes a single point and spins everything about that point. In three dimensions, a rotation fixes a line—the axis of rotation—and spins everything about that line. In two dimensions, a reflection is performed by using a line as if it were a mirror; in three dimensions, a reflection is performed by using a plane as if it were a mirror. These differences apart, though, the concepts are pretty much the same.

A cube, then, has at least eight symmetries— basically, the symmetries of that red face. Is this the lot? Not at all. The whole lure of the regular solids is that all faces are on the same footing. Choose any of the other five faces and color it blue. Rotate the cube to bring the blue face into the position originally occupied by the red face.

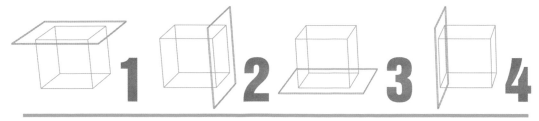

Having done so, there are four rotations and four reflections that keep the blue face in that new position. So now we've found eight more symmetries of the cube. Indeed, by the same argument, each face contributes eight symmetries. With six faces, that makes 48 symmetries altogether. That's an awful lot of symmetry.

Using the same kind of argument as for the cube, the octahedron also has 48 symmetries, the tetrahedron has 24, and both dodecahedron and icosahedron have 120.

Nature makes use of all these symmetries. Salt crystals are tiny cubes, quartz crystals can be octahedral; a molecule of methane is a tetrahedron, with a single carbon atom at its center and a hydrogen atom at each of its four corners. Many viruses, including chicken pox, are icosahedral—we'll see why later (see pp. 110–111). Famous drawings by Ernst Haeckel from the Challenger expedition of 1872–1876 depict radiolarians—microscopic marine organisms with siliceous skeletons—shaped like cubes, octahedra, and dodecahedra. However, it is suspected that he exaggerated their geometric regularity a bit.

The Pythagoreans associated the regular solids with the four elements: the tetrahedron with fire; the octahedron with air; the cube with earth; and the icosahedron with water. The dodecahedron they associated with the universe. They were wrong, but not totally wrong. In speculating that nature might employ highly symmetric structures, they were bang on target.

ABOVE The external form of a crystal is often symmetric. These symmetries are the visible trace of the deeper symmetries of the crystal's atomic lattice.

Crystallography was one of the areas of science that contributed to the mathematical theory of symmetry.

BELOW: LEFT TO RIGHT One of the simplest things that we can do with the mathematics of symmetry is to count how many symmetries a shape possesses. A cube, for example has 48 symmetries. The eight symmetries of a square—four rotations (1–4) and four reflections (5–8)—can be applied to each square face of the cube. Then, any of the six faces can be brought into the same position as that face. As a result, the cube has 8 x 6 = 48 symmetries.

 5 **6** 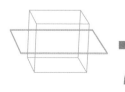 **7**  **8**

EARTHLY SPHERES

The regular solids have many, but a finite number of symmetries—24, 48, 120. Some three-dimensional shapes do better. A cylinder has an infinite number of symmetries—all rotations about its axis, all reflections in planes that contain this axis, and a top-bottom reflection as well. The most symmetric three-dimensional object of all—at least if we insist that the object should have a finite size—is the sphere. Just as the Greeks considered the circle to be the perfect form in two dimensions, so they considered the sphere to be the perfect form in three dimensions. And all this without even delving into the world of symmetry.

Because of the circle's symmetry, every point on its perimeter lies at the same distance from the center, so it can roll smoothly forward and backward, hence the wheel.

Because of the sphere's symmetry, every point on its surface lies at exactly the same distance from the center, so it can roll smoothly in any direction—which is why so many games use a round ball. Golf, basketball, cricket, tennis, baseball, soccer—it's all just symmetry, really.

What shape is a raindrop? Cartoonists have a habit of drawing rain in a teardrop shape, a blob with one rounded end and one sharp end trailing behind it. The shape caricatures the rapid motion of the falling raindrop, but it is no more accurate a representation of reality than the convention that cartoon characters' thoughts can be seen as words written on clouds that emanate from their heads. Human psychology leads us to expect a drop of falling rain to have this classic teardrop shape, but actually raindrops are spheres.

Well, as always, not quite. Air resistance can flatten the spheres, and in some circumstances

the spheres can vibrate. But for a tiny drop these effects are very small indeed—fine drizzle is a rain of damp spheres.

Why are raindrops spherical? Let's get rid of the air and let the drop fall in a vacuum. This eliminates distortions caused by air resistance. Surface tension pulls the drop into whichever shape has the least energy: fundamentally, nature is lazy. The energy of a liquid drop is proportional to its surface area, so the raindrop tries to make its surface area as small as it can. Its volume, though, is fixed by the amount of water it contains—for these purposes, water is incompressible.

What shape has the least surface area for a fixed volume?

According to ancient legend, Dido was given a bull's hide and told she could have as much land as the hide could surround. She cut it into thin strips and enclosed a circle large enough to found the city of Carthage. A circle is the shape of a given area with the shortest perimeter—or equivalently, the perimeter that encloses the greatest area, which is where Dido comes in. By analogy, it's no surprise to find that the shape of a given volume with the smallest surface area (or of a given surface area enclosing the largest volume) is a sphere. Experimentally this is pretty obvious, but proving it is distinctly tricky. However, it's been done, and the sphere is the right answer.

When the Earth first formed, it was a gigantic blob of molten rock and iron, mixed with various gases, steam, and all sorts of junk. As it circled the Sun in orbit, it was in free fall—centrifugal force and gravity nearly canceled out, that's what "orbit" means. So our planet, like the raindrop, was a blob of liquid in zero gravity. It therefore took up the same shape as the raindrop—a sphere. However, the proto-Earth was spinning. The forces generated by its rotation caused it to expand at the equator and flatten at the poles. The Earth's core is still molten. We live on the thin, solid crust. Convection currents caused by heat slowly move the Earth's interior around, like self-stirring custard, and the continental crust and the ocean floors move in response to this. The result is continental drift. Long ago, the continents were in quite different positions from where they are today—they slowly moved, and are still moving today.

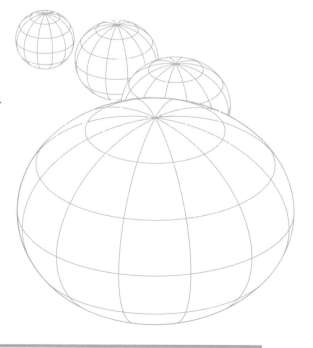

LEFT The sphere is the shape that has the smallest surface area while containing a given volume. Drops of water naturally pull themselves into spherical balls as surface tension makes them reduce their area.

RIGHT Planets, which at one stage are mostly molten rock, pull themselves into spherical shapes by the force of their own gravitational attraction. Rotation can cause the planet to flatten into an ellipsoid when still molten.

LEFT Gravity, the organizing force of the cosmos, causes every body to attract every other. The dramatic spiral of a galaxy is a consequence of the physics of gravity, which causes a random cloud of matter to collapse into a rotating disk, which later acquires the familiar spiral structure.

Theoretical models of these convection processes make good use of the Earth's spherical symmetry. Nonetheless, the continents are about as asymmetric as anything can be—but that's all right, we've already learned that symmetric causes can sometimes have asymmetric effects, for instance Jupiter's Great Red Spot. So a spherical planet can have continents the shape of Africa and Australia without violating any fundamental principles of the universe.

OTHER SOLAR SYSTEMS

The place where spheres really come into their own is in the cosmos. Planets, moons, and stars are all, to a first approximation, spherical. The spherical shape was important in the development of Newton's theory of gravity, because he was able to prove that the gravitational attraction exerted by a sphere is exactly the same as that exerted by a point concentration of matter with the same mass as the sphere. The proof assumes that the distribution of matter is spherically symmetric— at any given distance from the center, the density should be the same everywhere. This remarkable result let Newton replace spherical planets by point masses in his calculations,

making the calculations of orbits much simpler.

However, planets are only approximately spherical. Like the Earth, they tend to be slightly flattened as the result of their rotation—they are oblate ellipsoids. Their diameter at the equator is generally bigger than the distance from pole to pole. However, they come close, especially in the so-called terrestrial, or inner, planets to the spherical ideal. The equatorial and polar diameters of Mercury and Venus differ by less than one part in a thousand. The difference for Earth is three parts in a thousand, while for Mars it is seven parts in a thousand. The giant planets—Jupiter, Saturn, Uranus, and Neptune— have extensive atmospheres and small cores, and balls of gas deform more easily than molten rock, so it is no surprise to find that the giant planets depart further from the spherical. Saturn is the most oblate—the difference in its diameters is almost ten percent, clearly visible to the eye.

Since 1992 astronomers have discovered a large number of exoplanets—planets orbiting stars other than the Sun. By early 2016 there were 2,086 known exoplanets around 1,330 different stars, of which 509 stars have more than one planet. There are several ways to detect exoplanets. One is to measure tiny changes in the frequency of light emitted by the star as it

wobbles slightly under the gravitational influence of its planets. The commonest and most successful method is to observe changes in the star's light output when (and if) a planet passes in front of it. It's also possible, occasionally, to observe an exoplanet directly in a powerful telescope, using clever methods to stop the light from the star blotting out that from the planet.

These planets were not detected earlier because the measurements required could not be made with enough accuracy. Astronomers have long expected many stars, perhaps even most of them, to have planets. Our current theories of the formation of stars predict that planets will form around them as part of a single process of gravitational condensation. The process begins with a randomly fluctuating cloud of interstellar dust and gas. The fluctuations cause matter to concentrate in some region, and because gravity is a long-range force, this triggers a collapse of the whole cloud, with everything heading roughly toward a common center. Within a mere ten million years a dense dust cloud forms, roughly spherical in shape.

If left undisturbed, it would remain spherical, but it is part of a galaxy, and the galaxy is rotating. Galaxies rotate faster toward their centers, more slowly at their edges. So the part of the dust cloud that lies farthest from the galactic core starts to lag behind, while the opposite part, nearest to the galactic core, starts to move ahead. The result is to set the sphere of gas and dust spinning, which then breaks its spherical symmetry. However, the sphere is still continuing to collapse. The collapse is most rapid along the axis of spin, and slowest at right angles to the axis because there are centrifugal forces that counteract the contracting effect of gravity. The once spherical cloud then rapidly turns into a spinning disk—the spherical symmetry breaks to circular symmetry.

Near the middle, the disk thickens into a blob, and as the blob shrinks its density increases. Gravitational energy turns into heat and the temperature rises. When and if it gets hot enough, nuclear reactions are ignited and the blob becomes a star. Meanwhile, the rest of the disk has been obeying the general tendency of self-gravitating systems to form clumps. The clumps undergo their own local collapses and they too heat up, but they are too small to form stars. Instead, they form balls of molten rock, whose surface later cools—the star has planets. And many of the planets have moons, formed from even smaller clumps.

BUBBLE, BUBBLE

Celestial spheres are impressive, but the best mathematics often takes its inspiration from simple things. What could be simpler than a bubble? The mathematics of bubbles first got going in the 1830s, when the Belgian physicist Joseph Plateau began dipping wire frames into soap solution and was astounded by the results.

When he dipped a cubic frame into the soapy mixture, for instance, the soap films arranged themselves into 13 almost flat but slightly curved faces, like 12 triangles with their tips cut off, meeting in the center of the cube at the edges of a small square. A double loop of wire led to a Möbius band of soap—a surface with only one side. A Möbius band can be made by taking a strip of paper, giving it a half twist and then joining the ends—the soap film makes the same surface in one go, and the twist is imparted by the wire loop.

Despite 180 years of research, many of Plateau's observations still lack rigorous explanations. An especially notorious case was

that of the Double Bubble Conjecture, which describes the shape formed when two spherical bubbles coalesce. Plateau noticed that when two bubbles stick together they appear to form three spherical surfaces. The bubbles meet along a circular boundary and the interface—itself a portion of a sphere—bends a little into the bigger bubble. Area-minimization implies certain restrictions on the sizes of the spheres and the angles between them. The challenge to mathematicians is to prove that no other two-bubble shape has smaller area.

The first step is to idealize the problem. A soap bubble is a minimal surface—a surface whose area is as small as possible, subject to suitable constraints. Bubbles form minimal surfaces because the energy in a soap film is proportional to its area. For example the minimal surface that encloses a given volume is a sphere, and that's why individual soap bubbles are spherical. Minimal surfaces are of central importance in mathematics, with many applications including biology, chemistry, crystallography, and architecture. Bubbles form foams, so there are

applications to brewing, too, and to the packaging industry, which uses slabs of plastic foam to protect goods from damage during transit.

The first major success confirmed Plateau's experimental observations of the angles where soap films meet. He found that they either meet in threes, at angles of 120 degrees, or in fours, at angles of 109 degrees. These two angles are fundamental to any problem in which soap films abut each other, and in 1976 the American mathematicians Jean Taylor and Fred Almgren proved conclusively that in all minimal surfaces these two angles, and no others, occur. Their proof comes in two parts. First, they show that all arrangements can be reduced to a list of ten candidate configurations of surfaces; then they eliminate all but two by proving that in each of the other eight cases the total area can be made smaller by changing the configuration.

It is easy to prove the Double Bubble Conjecture if you assume at the outset that the bubbles are parts of spheres. The hard part is to prove that there are no smaller-area alternatives.

That's not so obvious—for example one bubble might form a doughnut and the second might fit through it like a dumbbell.

In 1995 the mathematicians Joel Hass and Roger Schlafly found a way to rule out such bizarre alternatives for equal bubbles. Their proof was computer assisted—they had to work out 200,260 different integrals. Such is the speed of modern computers that this task took a mere 20 minutes. A proof that can deal with the general case, unequal bubbles, was found in 2000 by another group of four mathematicians. This handles far more possibilities than any computer could ever hope to calculate, yet was calculated using no technology more advanced than pencil and paper.

The subject has already moved on—a team of three undergraduates has extended the result to four-dimensional spaces, with inroads into five dimensions. In this world of computers, it is reassuring that the human brain can still out-think the machine.

LEFT AND BELOW The mathematics of bubbles explains what shapes they will form under given conditions. When a wire cube is dipped in soap, the bubbles form thirteen almost-flat surfaces (1). A double loop of wire produces a Möbius band (2). The form of a double bubble can be predicted, on the assumption that surfaces involved are spherical (3). And it has now been proved that these surfaces must be spherical, because alternative shapes such as a torus and dumbbell can all be ruled out (4).

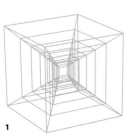

1

2

3

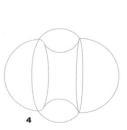

4

DOMES & VIRUSES

If you want to make a surface containing the largest amount of space within the smallest possible area, then the answer is a sphere. Sometimes, however, a sphere is ruled out as a solution because the surface has to be made from a number of rigid units. This extra constraint occurs in the microscopic world of viruses, where the surface of the virus has to be made from a lot of identical protein units, and in architecture, where an approximately spherical dome has to be made from flat panes of glass. Both nature and the architect Buckminster Fuller hit on the same solution—make whatever shape you can manage that is closest to a perfect sphere. All such shapes are based on the icosahedron, the most nearly spherical of the five regular solids.

An icosahedron has 20 triangular faces. If you cut off all the corners, so that exactly one third of each triangular edge remains, you get the truncated icosahedron, with 20 hexagonal faces and 12 pentagonal ones. This shape is commonly used to make soccer balls. It is strong, almost spherical, and made from flat pieces, which bend slightly when the ball is inflated, making it even closer to a true sphere.

In 1750 Leonhard Euler, a Swiss-born mathematician, proved a fundamental relationship that governs such arrangements.

(René Descartes was aware of the result in 1639 but did not publish a proof.) Consider any "simply connected" polygon—one that can be continuously deformed into a sphere. Euler showed that the number of faces plus the number of vertices is always equal to the number of edges plus two. For example a cube is simply connected—imagine making a hollow rubber cube and blowing it up like a balloon so that its faces bulge to form a sphere. A cube has 6 faces, 8 vertices, and 12 edges, and 6+8 = 12+2, in accordance with Euler's result. (Nonsimply connected polyhedrons—an empty picture frame is an example—also exist, and for these Euler's relationship has to be modified.) A clever calculation based on Euler's theorem proves that any solid made from faces that are either pentagonal or hexagonal must have exactly 12 pentagons. For instance this is what happens in the truncated icosahedron.

There is more freedom as regards the number of hexagons. The truncated icosahedron is the simplest of an infinite family of solids, known as pseudo-icosahedra. These have 12 pentagonal faces, 20T-12 hexagonal faces, 30T edges, and 10T+2 vertices, where T is any number of the form a^2+ab+b^2.

The architect Buckminster Fuller suggested using such polyhedra to make geodesic domes. The hexagonal faces are subdivided into six

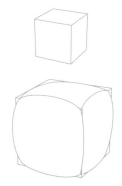

RIGHT Whoever would have believed that a humble soccer ball with its five- and six-sided panels would be material for a Nobel prize? All such arrangements must obey a simple mathematical rule, provided—like the cube—they can be deformed into spheres.

equilateral triangles, and the pentagonal ones into five almost-equilateral triangles—the eye can hardly tell the difference, so the dome looks as if it is made from lots of identical pieces. The best known geodesic dome, the United States Pavilion at Expo 67 in Montreal, was a pseudo-icosahedron with a = 16, b = 0.

The same shapes occur in many viruses because they are the best way to pack identical units while minimizing energy. The outer coat of a virus is typically made from many copies of the same protein unit, and these fit together like the vertices of a polyhedron. Turnip yellow mosaic virus is a pseudo-icosahedron with a = 1, b = 1. The rabbit papilloma virus has a = 2, b = 1. The chicken pox virus has a = 4, b = 0. In general the number of protein units is given by the "magic number" 10(a2+ab+b2)+2, and the magic numbers up to 300 are: 12, 32, 42, 72, 92, 122, 132, 162, 192, 212, 252, 272, 282. These—and no other numbers in that range—are the numbers of protein units that you should expect to find within a virus.

The truncated icosahedron also arises in the remarkable molecule buckminster-fullerene ("buckyball"), made from 60 carbon atoms. It is an entirely new form of carbon, first synthesized in 1985 in a collaboration between the Nobel prize-winning English spectroscopist Harry Kroto and the American chemist Richard Smalley. On September 1 of that year they vaporized carbon in an atmosphere of hydrogen, nitrogen, and various other elements to simulate the conditions near red giant stars, where this form of carbon was thought perhaps to exist. On September 4 they detected the presence of carbon molecules with molecular weight 720. Carbon's molecular weight is 12, so this value corresponds to exactly 60 carbon atoms.

What is the structure of the new molecule? The two scientists tried all sorts of ideas. Their graduate students discovered that the molecule was so stable that it could not have any "dangling bonds." This reinforced a feeling that it was a kind of polyhedral cage. Smalley recalls sitting up all night on September 9 with scissors and paper, and finding a possible structure—a truncated icosahedron. Variants of this form, collectively known as fullerenes, are currently the subject of intense research because they are a potential source of new materials for engineering and technology.

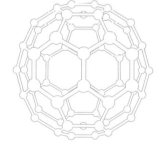

LEFT AND ABOVE These rules also govern the form of geodesic domes (left) invented by the architect Buckminster Fuller. The fullerene molecule is a cage made from 60 carbon atoms, shown above. Its discoverers won a Nobel prize. And what is the complete structure of the fullerene molecule? The same as that of the soccer ball.

SCREW THREADS & HELICES

In two dimensions, or even in the "essentially" one-dimensional pattern of a frieze, the interesting possibility of a glide reflection arises. This exotic symmetry is a combination of a reflection and a translation in a special direction—along the line of the mirror. In three dimensions there is an analogous possibility, in which a rotation is combined with translation in a special direction—along the axis of rotation. Such a symmetry is called a screw.

The name is appropriate. The reason that a corkscrew can enter a cork without doing too much damage is that the corkscrew has screw symmetry. Indeed it has an infinite family of screw symmetries. Whichever angle you turn it through, there is a corresponding distance through which it can be translated so that it still fits the same "spiral" hole in the cork. In fact that distance is proportional to the amount of turn. Nuts fit on bolts for the same reason.

The spiral curve involved is more correctly called a helix. It differs from an ordinary spiral by existing in three dimensions rather than two. Bolts have helical threads, and a corkscrew

has a helical prong. Wood screws also have helical threads, but they are tapered to ensure a good grip.

There are two kinds of helix, one with left- and one with right-handed threads. To appreciate the distinction, try to uncork a bottle of wine with a corkscrew designed for left-handed people (or vice versa). An ordinary right-handed corkscrew turns in a clockwise direction as it enters a cork: its mirror-image left-hand screw turns counterclockwise.

Nature makes good use of this distinction. Many climbing plants put out helical tendrils to secure themselves to walls, trellises, or other plants. One problem nature must solve is how to tighten up such a coil when its ends are firmly fixed. It does so by using a trick that mathematicians, rather unfairly, call perversion. At some point along its length the coil switches direction from right to left. The resultant kink can rotate, tightening or loosening the coil, without affecting the ends. Telephone cables often tangle into a perverted state.

After the pseudo-icosahedra, the next commonest form for a virus is the helix. A good example is the tobacco mosaic virus, made from

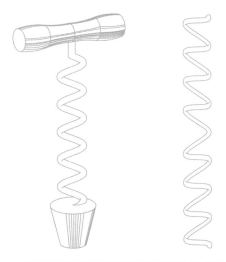

LEFT Like DNA the business end of a corkscrew is a helix, which possesses screw symmetry (far left). Recall that helices can occur in two forms, left- and right-handed, as do corkscrews. Many climbing plants use helical tendrils to secure themselves to walls or other plants, and they too can be left- or right- handed. Often, though, plants use helices with both orientations within the same tendril, changing the direction in mid-stream (left). With this arrangement, the tendril can easily be tightened up from the middle without disturbing the ends.

2,130 identical protein units fitted together like the steps of a spiral staircase. The symmetry here is discrete rather than continuous—rotations through specific angles must be combined with translations through specific distances. This restriction occurs because the resulting screw symmetry must move a protein unit into a precise coincidence with some other protein unit.

In architecture, the helix is used to create spiral staircases. The Château de Chambord in the Loire Valley has a double spiral staircase, in which two independent helices are intertwined—one for nobility, the other for servants. A similar double helix has become an icon of 20th-century science, and will form the basis of much of the science of the 21st century. This is the molecule DNA (deoxyribonucleic acid). DNA carries the "genetic information" of most living organisms. Although some viruses use the closely related molecule RNA instead. The "steps" of the DNA staircase are matched pairs of molecules—adenine/thymine or cytosine/guanine. The genetic information is written in a four-letter code: A, T, C, or G. Some regions, called genes, specify the sequence of units from which proteins necessary for life are made—three code letters per unit. The units are molecules called amino acids.

There are 64 triplets but only 22 amino acids (as well as an "empty" amino acid indicating "stop"), and this redundancy is related to a kind of symmetry in the genetic code that translates the triplets of code letters into amino acids. This symmetry is highly imperfect. Its most obvious manifestation is that often four different triplets code for the same amino acid. (It can also be up to six, or just one.) When it is four, the third letter can be changed without altering the unit—a permutational symmetry of the code. Perhaps at some stage in Earth's long history, life was simpler and used only a two-letter code.

ABOVE Spiral staircases are also helical, like the above stone staircase, which was built in a French château.

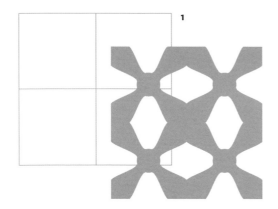

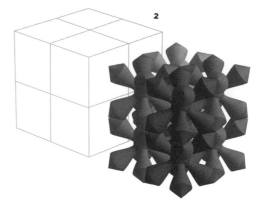

CRYSTAL LATTICES

Viruses are symmetric because they are made by packing identical protein units efficiently. This brings us back to Kepler and his idea that ice crystals are formed by packing identical units. He thought the units were globes of vapor, which is wrong, but that's a minor quibble. How do the basic units of a crystal pack? The analogous problem in two dimensions, that of tilings, leads to the 17 wallpaper symmetries (*see pp. 78–79*). What of three-dimensional wallpaper?

The mathematical essence of a wallpaper pattern is its symmetry, and the key feature here is that the pattern repeats the same element regularly in two different directions. The pattern is arranged according to a plane lattice. The atoms in a crystal are arranged in a three-dimensional lattice. It is the lattice structure that creates the regular geometry of crystals and, in particular, places the restrictions on the angles between their faces, which so intrigued the early crystallographers.

The geometric possibilities in three-dimensional space are richer than they are in two dimensions, and it is not surprising to discover that this leads to a wide variety of crystal symmetries. The 17 symmetry types of wallpaper pattern are trumped by a massive 230 symmetry types of crystal lattice.

The symmetries of a crystal lattice come in two kinds. First, there are the lattice translations, which provide a kind of skeleton on which to hang a repeating "design." Then there are the symmetries of that arrangement itself. The same division also applies to wallpaper. For example if we start with a square lattice, a square grid of identical tiles, then we can decorate the tiles in different ways. We could use a design with the full symmetry of the square, say, or just with its rotational symmetries. In this way we get two distinct symmetry types of wallpaper pattern, based on the same symmetry type of lattice.

The problem of classifying symmetry types therefore splits into two parts. First consider just the lattices, then decorate them with symmetric "designs"—which, in the crystal case, are arrangements of atoms. There are five types of two-dimensional lattice, based on a parallelogram, rectangle, rhombus, square, or hexagon. There are 14 symmetry types of three-dimensional lattice, known as Bravais lattices. For each type, a separate analysis of the possible types of "design" must be carried out. Put it all together, and you emerge with those 230 possibilities.

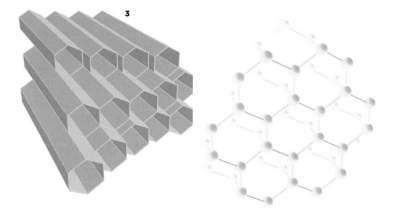

3

LEFT Planar lattices can be constructed by fitting together many copies of the same basic unit (1). The same goes for lattices in space, but now there are far more possibilities—230 in all (2). Among them is the atomic lattice of ice, which resembles a stack of hexagonal prisms (3). The sixfold symmetry of this lattice is responsible for the sixfold symmetry of snowflakes.

All this bears on the question of the snowflake. What does a snowflake pack together? Water molecules. Ice is crystalline water. It comes in many forms, but the commonest one is based on a crystal lattice that is a slight variation on the hexagonal structure of the honeycomb. This lattice forms at normal atmospheric pressure and a temperature just below 32°F (0°C). At lower temperatures and higher pressures, the crystal structure of ice can be different.

A water molecule is a tetrahedron with an oxygen molecule at its center. Two vertices of the tetrahedron are occupied by hydrogen atoms and the other two are empty. Ice crystals stack these tetrahedra together in a regular pattern. The lattice of ice, the one relevant to snowflakes since they form at "normal" temperatures and pressures, consists—approximately—of stacked layers of hexagonal prisms. The oxygen atoms lie at the corners of the component hexagons, and the hydrogen atoms lie about one-third of the way along the edges.

When these layers are viewed square on you see long hexagonal tunnels like a honeycomb. The layers are stacked above each other without any sideways displacement of the hexagons and the hexagonal ends of the prisms abut precisely. If you look at the lattice sideways on, however, it becomes clear that the hexagonal layers are

(nearly) flat. Although they do have slight dimples, up and down—the ends of the prisms are bumpy. These dimples are what Kepler got wrong; the basic honeycomb is what he got right.

The flat layers can slide relatively easily over each other, which is one reason why ice is slippery. Ice crystals also grow faster along these flat layers, because there are two places to attach a hydrogen atom within a layer but only one place at right angles to the layer. This is why an ice crystal starts as a flat hexagonal plate— the regular seed from which its decorative fernlike beauty grows.

THE KEPLER CONJECTURE

About one third of the way through his book, *On the Six-Cornered Snowflake*, Kepler made a statement that caused the world's mathematicians and physicists nearly 400 years of grief. He was discussing ways to pack identical pellets in three dimensions. He starts with what we would now call a cubic lattice: "The arrangement will be cubic, and the pellets, when subjected to pressure, will become cubes. But this will not be the tightest pack. In the second mode, not only is every pellet touched by its four neighbors in the same plane, but also by four in the plane above and by four below, and so throughout one will be touched by twelve, and under pressure spherical pellets will become rhomboid… The packing will be the tightest possible, so that in no other arrangement could more pellets be stuffed into the same container."

Although Kepler talks in terms of a container, he does not specify any particular shape or size for it. Since mathematicians define the efficiency of a packing of infinite space by considering what happens inside larger and larger finite regions, we may as well assume that Kepler is talking about filling the whole of space. In this form, his statement rapidly came to be one that "most mathematicians believe and all physicists know."

The arrangement Kepler describes is what we now call the face-centered cubic lattice—the lattice points are the vertices of a stack of cubes, together with their centers. It is the arrangement commonly used by greengrocers to stack oranges. It might seem obvious that this has to be the most efficient packing in three dimensions, but such things are hard to prove. We saw that the analogous question in two dimensions, with the obvious honeycomb answer, was not resolved until early in the 20th century (see pp. 80–81). The face-centered cubic lattice packs spheres to a density of 74 percent. The most obvious rivals are the hexagonal lattice, which manages 60 percent, and the ordinary cubic lattice at 52 percent. It's easy to compare these numbers and see that the face-centered cubic lattice wins, but that's not the difficulty. How can we be sure that the optimal packing is a lattice? This question opens a whole can of worms.

The Kepler Conjecture, as the mathematical mystic's innocent remark came to be called, resisted proof right up until a few years ago. In 1994 the American mathematician Thomas Hales proposed a five-step program to prove the conjecture. It involved finding a useful

representation of the geometry of all spheres "near" a given one. Hales decided that two spheres would be considered "near" if the distance between their centers is less than or equal to 2.51 diameters. Given the sphere-packing pattern, we can form a network by drawing lines between the centers of any two spheres that are near each other using this definition. This network forms a kind of skeleton of the packing and allows us to make sense of the geometry of the packing near any given sphere. In essence, we have to prove that in an optimal packing, this local geometry is the same as that of the face-centered cubic lattice.

In 1953 the Hungarian mathematician Fejes Tóth used a version of this idea to reduce the Kepler Conjecture to a specific calculation, which was far too big for anyone to carry out, even with the aid of computers far more powerful than any that exist today. Very roughly, the idea is to consider any local geometry that is not the one found in the face-centered cubic lattice and show that such an arrangement can be changed to increase the efficiency of the packing.

Hales modified Tóth's approach, replacing the geometry of curved multidimensional surfaces by the simpler geometry of multidimensional planes. This increased the number of different cases that had to be tackled, but made each case simpler. The final description occupied 250 pages and is supported by 3 gigabytes of computer code and data. Because the proof is too long to be checked by a human, Hales set up the "Flyspeck Project" to produce a new proof that could be verified rigorously by a computer. The project was completed successfully in 2014.

I was rather hoping that it would turn out that Kepler was wrong. Kepler, along with all those physicists who think that the answer is so obvious there's no need for a proof. No such luck. Kepler's intuition—and that of the physicists— is unfortunately fully vindicated.

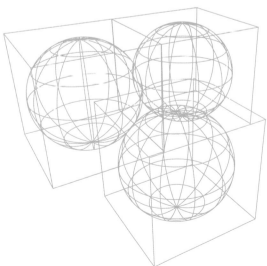

ABOVE AND LEFT What then is the most efficient way to pack spherical fruit—that is, what arrangement will leave us with the smallest gaps? Kepler convinced himself that the best way is the method that every greengrocer uses. The difficulty, as we have seen, is not to guess the right answer, but to find a logically watertight proof. Might there, perhaps, be some other cunning arrangement that packs the fruit a little bit more tightly than the usual one? How can a person decide, if you don't know what it is?

10

SCALE & SPIRALS

Symmetries can also make things change size. Scientists talk an awful lot about scales. Not the kind you weigh things with, or the kind you find on fish, but scales of space and time. Big/small, fast/slow, that sort of scale. Maps represent territory effectively because they have the same shape, but on a different scale; toy models of cars and aircraft reproduce the shape of the original, but on a smaller scale.

A symmetry that changes the scale of an object or system is called a dilation. A dilation multiplies all distances by a fixed amount, the scale factor. If the scale factor is smaller than one, all the distances shrink and we have a contraction. If the scale factor is greater than one, all the distances grow causing the object to become enlarged and expanded.

Different physical properties of an object behave in different ways when the object is dilated. The simplest behavior, for instance,

is that of length. If you make an object twice as big (dilate using the scale factor 2) then its length (also breadth, width, height, waist measurement—if it has one—and any other linear distance that makes sense) also becomes twice as big. The same goes for any other multiple.

Areas are different. If you double an object's size, then its area (surface area, cross-sectional area in a particular position, and any other area that makes sense) is multiplied by four. Triple its size, and the area multiplies by nine. So the area scales like the square of the length—that is, you have to multiply the scale factor by itself.

Volumes have yet another scaling behavior. Volume scales in the same way as the cube of the length—multiply the scale factor by itself, and then multiply the result by the scale factor again. Double the length, and the volume is multiplied by 2x2x2 = 8. Triple the length, and the volume is multiplied by 3x3x3 = 27.

Relations of this type are called scaling laws. These relationships represent a kind of dilational symmetry which occurs in certain laws of nature, or in concepts that relate to those laws.

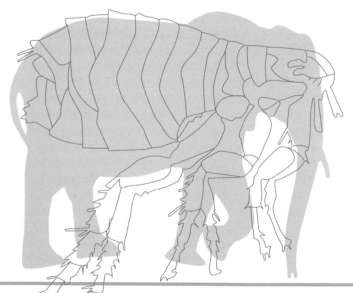

LEFT AND ABOVE RIGHT If a flea were the size of an elephant (left), would it be able to jump over the Eiffel Tower? The mass of an animal, and the strength of its muscles, scale differently according to the animal's size. Mass scales as the cube of the animal's size, but muscle strength scales as the square of the size (above right). The flea's spindly legs are fine when it's flea-sized, but they would never support an elephant.

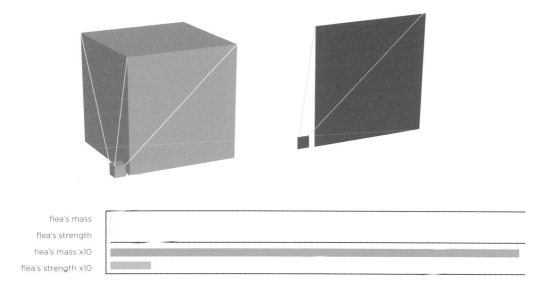

flea's mass	
flea's strength	
flea's mass x10	
flea's strength x10	

Dendritic snowflakes obey scaling laws. Each small "twig" is an exact miniature version of a larger branch. So do some living creatures, their forms reflect nature's scaling laws—with some degree of freedom for maneuver because biology is good at bending the rules. Life is amazingly flexible. It often seems to be at odds with the laws of physics. But it can't break them. This can be a source of confusion. Graphic descriptions of the abilities of living creatures often dramatize them through a change of scale. For example "If a flea were the size of an elephant, it would be able to jump over the Eiffel Tower." That kind of thing.

Really?

OK, we understand what such a statement is telling us. It's a metaphor. A flea is about the size of a grain of sugar. Yet it can jump as high as a bag of sugar. The numbers are boring, sugar is humdrum—stated with elephants and Eiffel Towers, it's more impressive. However, let's take the metaphor literally. If we somehow made a jumbo-sized flea, how high could it jump?

The height to which an object rises depends on two things—its mass and the force that propels it upward. The height is proportional to the force, but inversely proportional to the mass (big masses jump to small heights). Mass is proportional to volume, so it scales as the cube of the size. How does a flea's jumping-force scale? That's a difficult question to answer accurately, but we can get a reasonable idea by observing that the jumping force is exerted by muscles, and for a ballpark figure the most important feature of a muscle is its cross-sectional area. This, we know, scales as the square of the size.

Cubes are bigger than squares, so the scaling of the mass wins by comparison with muscular cross section. Conclusion—a flea the size of an elephant would hardly jump at all. Indeed, an animal's ability to support its own weight also depends on a cross-sectional area (of its skeleton, if it has one—an internal bone skeleton for an elephant, an external chitin one for a flea). So a flea the size of an elephant would be unable to stand, let alone jump. Its legs would break under the weight of its body.

Low note

High note

MATHEMATICS OF MUSIC

Say the word "scale" and music immediately springs to mind, but there the term has its own special meaning. Nonetheless, musical scales have their own dilational symmetry. If you look at a guitar, mandolin, or lute—any stringed instrument with frets—you'll see that the frets get closer and closer together as the note gets higher. This is a consequence of the physics of vibrating strings. Today's Western music is based upon a scale of notes, generally referred to by the letters A to G, together with symbols $^\sharp$ (sharp) and $^\flat$ (flat). Starting from C, for example successive notes are

C$^\sharp$ D$^\sharp$ F$^\sharp$ G$^\sharp$ A$^\sharp$

C   D   E   F   G   A   B

D$^\flat$ E$^\flat$ G$^\flat$ A$^\flat$ B$^\flat$

and then it all repeats with C, but one octave higher. On a piano the white keys are C D E F G A B, and the black keys are the sharps and flats.

This system can be traced back to the Pythagoreans, who discovered that the intervals between harmonious musical notes can be represented by whole number ratios. They demonstrated this experimentally using a single string. The most basic such interval is the octave. On a piano it is a gap of eight white notes; on the Pythagoreans' string it was the interval between the note played by a full string and that played by one of exactly half the length. The ratio of the length of string that produces a given note to the length that produces its octave is 2:1.

ABOVE The sounds made by a vibrating string or drumskin depend on its shape and size. A short string produces a higher note than a long one (see above). The spacing of frets on a guitar ensures that each successive fret moves the pitch one note up the musical scale. The spacing is wide for the low notes and narrow for the high ones.

B C D E F G A

The Pythagoreans discovered that other whole number ratios produce harmonious intervals as well. The main ones are the fourth, a ratio of 4:3, and the fifth, a ratio of 3:2. It is thought that, in order to create a harmonious scale, the Pythagoreans began at a base note and ascended in fifths. This yields a series of notes played by strings whose lengths are successive powers of 3/2:

1 3/2 9/4 27/8 81/16 243/32

Most of these notes lie outside a single octave, that is, the fractions are greater than 2. We can descend from them in octaves, dividing successively by 2, until the fractions lie between 1 and 2. Rearranging in numerical order, we get:

1 9/8 81/64 3/2 27/16 243/128

When played on a piano, these correspond approximately to the notes C D E G A B.

Strangely, the gap between the ratios 81/64 and 3/2 sounds "bigger" than the others, and it can be plugged by adding another, the fourth, a ratio of 4/3, which corresponds to F on the piano. When this is done, the intervals between successive notes form two distinct fractions: the tone 9/8 and the semitone 256/243. An interval of two semitones is very nearly, but not exactly, equal to a tone. Thus there are gaps in the scale: each tone must be divided up into two intervals, and an octave then contains 12 semitones.

There are various schemes for doing this, but the only way to make all 12 intervals equal requires the semitone ratio to be the twelfth root of 2, roughly 1.05946. The resulting scale is said to be equitempered. The frets on a guitar follow the same spacings—each gap between successive frets is the previous gap divided by the twelfth root of two. Thus the guitar fingerboard has a dilational symmetry, with scale factor 1.05946.

Many mathematical problems have been inspired by music. In 1966 the American mathematician Mark Kac asked: "Can you hear the shape of a drum?" A more impressive version

is: what information about a shape can you infer from the frequencies with which it vibrates? This question has practical significance. For instance when an earthquake hits, the Earth rings like a bell and seismologists deduce a great deal about the internal structure of our planet from the "sound" that it produces.

Kac showed that some features of a drum are determined by its sound, for example its area and its perimeter. In 1992 some of his compatriots, the mathematicians Carolyn Gordon, David Webb, and Scott Wolpert, constructed two distinct mathematical drum skins that produce the identical range of sounds. They are made by fitting together pieces shaped like half a Maltese cross in two different arrangements. Not only can one drum be cut-and-pasted to give the other, but the vibrational patterns can be cut-and-pasted as well. So to any vibration of the first drum there corresponds a vibration of the other drum that makes exactly the same sound.

BELOW It has been known for over 50 years that if two drums have different areas, or different perimeters, then they produce different sounds. Surprisingly, though, two drums with different shapes can produce exactly the same sounds.

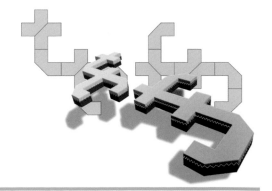

SPIRAL GROWTH

Scaling laws and musical scales are rather abstract examples of dilational symmetry. Are there examples where we can actually see the pattern of dilation, in the same way that the snowflake lets us see rotational and reflectional symmetries? There certainly are, especially if the dilation is combined with a rotation. The pattern that results is a spiral. Nature makes extensive use of spirals.

A spiral is a curve that winds around and around a central point, getting further and further from it in one direction, and closer and closer to it in the other. There are many kinds of spiral, but only one kind has exact dilational symmetry. This is called a logarithmic spiral. The name arises because the angle through which it turns is given by the logarithm of the radius. A (slightly) friendlier way to describe it is to imagine an infinitely long rod, rotating about a fixed pivot at a constant speed. Now imagine a pencil moving along the rod, away from the pivot, at a speed that grows faster and faster. The speed should double over a particular period of time. As the rod whirls and the pencil whizzes away, the tip of the pencil will draw a logarithmic spiral.

The most obvious qualitative feature of a logarithmic spiral is that it is very tightly wound near the center, but successive turns get further and further apart as the distance from the center increases. A quantitative explanation of this pattern of winding is that the curve is symmetric under a particular combination of a rotation and a dilation. In fact, any rotation of the curve can be combined with a suitable dilation, so that the result leaves the form and position of the spiral apparently unchanged.

The best known natural example of a logarithmic spiral is the shell of the Nautilus, a marine mollusk, indeed a cephalopod, found in

the deeper parts of the South Pacific and Indian oceans. The animal itself has long tentacles, which it uses to catch crabs for food. Its shell is an elegant logarithmic spiral, divided into chambers of increasing size. The shape of the Nautilus's shell can be explained by its dilation-rotation symmetry, which arises from the creature's pattern of growth. The Nautilus is soft, but its shell is hard. A growing Nautilus cannot expand itself inside a fixed shell; neither can its shell expand to accommodate a larger inhabitant—unless the inhabitant builds an extension onto its house. And this is just what the Nautilus does. It continually adds material to the edge of its shell, and as the Nautilus grows exponentially, so does its shell. All of which leads, inevitably, to a logarithmic spiral. Of course, a mathematician's

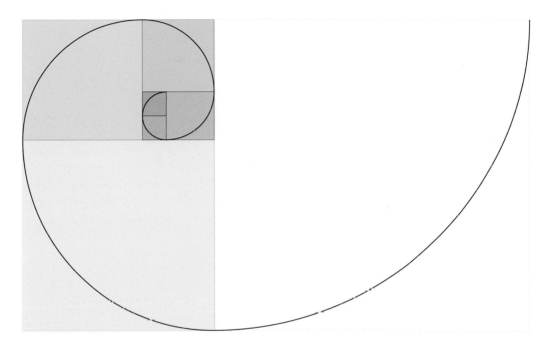

logarithmic spiral is infinitely big, but a Nautilus shell isn't—another case of mathematics idealizing biology.

Measurements show that in the Nautilus, each successive coil is roughly three times the width of the previous one. In other spiral shells this ratio can be different—apparently, there is no universal value. D'Arcy Thompson quotes figures for over 40 types of shell, in which the ratio of each coil to the previous one ranges from 1.14 to 10.

As mentioned earlier (*see pp. 22–23*), other well known spiral shells are those of the extinct ammonites—creatures that lived roughly 300 million years ago in the Devonian, Carboniferous, and Permian seas and are widely found as fossils. Some ammonite shells have such a small growth rate that they look more like an Archimedean spiral, in which successive coils are separated by the same amount (a ratio of 1). However, most ammonites have logarithmic spiral shells. And other shells have all sorts of weird spiral shapes. So the main message here is that the form of the shell is some kind of record of the way its inhabitant grows and the rate at which it builds new shell material. The pattern that we see in a shell is a clue to the animal's rules of growth.

Aha!

The pattern that we see in a snowflake can provide us with a clue to the snowflake's rules of growth.

I just knew this spiral stuff would come in handy.

SEASHELL FORM & PATTERN

As well as the shapes of shells, mathematicians have studied their patterns. This is an attractive problem because the shell is essentially a surface, and growth occurs only along its edge. Once part of the pattern has formed, it doesn't change. So at any stage in the process, a single new line of shell pattern is being added to whatever pattern already exists. This implies that the pattern is a space-time diagram of the chemical dynamics of the pigments that have been deposited, with the time direction being the direction in which the shell rolls up. Mathematically, that's neat and elegant.

The shell grows because the creature that inhabits it deposits minerals excreted by its mantle. The most extensive work on the patterns of seashells has been done by the German mathematical biologist Hans Meinhardt, and his starting point is Turing's theory of chemicals that react with each other and diffuse through some substrate (*see pp. 94–95*). The combination, as we have seen, creates elegant spatial patterns. Meinhardt's idea is to work backward from the patterns of seashells to find what kind of chemistry gives rise to them. In this way, he can infer certain features of the underlying biology.

Shell patterns are varied, but they fall into a number of basic types. These include regular stripes and spots, wavy stripes, and curious semiregular patterns involving triangles and zigzag lines. Two apparently different patterns can alternate, for example a pattern of dark spots may be interrupted every so often by a pale patch without spots.

Meinhardt argues that most shell patterns are created by a chemical scheme involving short-range enhancement and long-range inhibition. The production of pigment in a region causes other regions to produce pigment—but not necessarily the same one. At the same time it suppresses distant pigment. The creative tension between the short-range activator and the

long-range inhibitor leads to pattern formation, but the inhibitor needs a competitive edge for the process to work properly. In fact, a necessary condition for a pattern to form is that the inhibitor should diffuse seven times as fast as the activator.

The angular triangles and zigzag patterns can be viewed as the space-time diagrams of traveling waves. A triangle, for instance, forms when pigment is deposited at some isolated point, a pacemaker region. As time passes, waves of pigment production travel away from the pacemaker region—one wave moves to the left, the other to the right. The result is a triangle. If there are several pacemaker regions, adjacent triangles eventually run into each other, and what they do then depends on the chemical rules that are in play. A common outcome is that the two waves annihilate each other and disappear— the resulting pattern manifests itself as a zigzag.

Oscillatory dynamics, periodic in time, lead to shell patterns that are periodic in space. Stripes,

spots, and similar patterns arise easily from this mechanism. A particularly interesting possibility, to which I shall return when we have developed the necessary background (*see pp. 164–165*), is found in shells like the textile cone and the olive shell. These patterns display a mixture of well-defined basic shapes—spots, triangles, scales—but arranged somewhat randomly. It may seem surprising that simple mathematical rules can generate something so complex, but computer simulations show that this is actually a common feature of the Turinglike systems that Meinhardt uses.

The most complex patterns, where different patches of the shell have different kinds of markings, result from the interaction of two distinct systems, each involving different chemicals and laying down different pigments. The rules of interaction say that under some circumstances one system will win, whereas under other circumstances it will lose. In patches where that system wins, it then makes its own particular pattern; when it loses, we then see the alternative pattern. By thinking like this, it is possible to "reverse engineer" the chemistry from the patterns.

Although real shells are not flat, they are built by growth along the edge of a surface, and the patterns are largely independent of the shape into which that surface rolls itself. Mathematical models of a different kind from Meinhardt's— sophisticated versions of our observations about the growth of the Nautilus—can be used to investigate how different growth processes lead to different shapes. Przemyslav Prusinkiewicz and colleagues have been especially active in this area. By combining their methods with Meinhardt's it is possible to produce convincing shells in three dimensions, complete with markings, from simple mathematical recipes. Nature must play very similar games.

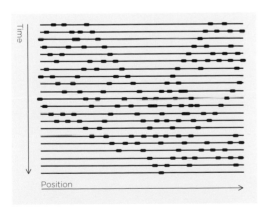

Time

Position

LEFT AND ABOVE Each new layer of pigment in the olive shell (left) depends on the previous layer, in a way that causes the spots of dark pigment to move one step sideways at each stage.

This results in a pattern of sloping lines, which fit together to form triangles (above). The markings of the olive shell look irregular, but are a consequence of simple mathematical rules.

PLANT NUMEROLOGY

Patterns of growth lead us back to plants and flowers. Recall that Leonardo of Pisa, more familiarly known as Fibonacci, invented his famous sequence 1, 2, 3, 5, 8, 13… (in which, apart from the first two numbers, each number equals the sum of the two preceding numbers) as a problem in an arithmetic text. We now have a fairly complete understanding of how the dynamics of plant growth leads to Fibonacci numbers. The detailed biological mechanisms behind the dynamics are less well understood, however. For instance we think certain hormones are used to suppress growth in particular places, but we aren't sure which hormones.

Why Fibonacci numbers? The short explanation is that the way plants grow leads to a preference for a small range of geometries, the most interesting being helical patterns based on the so-called golden angle. This angle, roughly 137.5 degrees, bears a strong mathematical relationship to the Fibonacci numbers and is responsible for their appearance.

The longer explanation is… longer. When a young plant emerges from the soil and starts to grow, the main source of activity can be found in the tip of the shoot. Here, cells are constantly dividing to create new cells. The cells migrate from the tip of the shoot to its edge, and as they do so, patterns of genetic and biochemical activity set the scene for the later development of sideshoots, petals, seeds, and all the other plant organs.

Close to the growing tip, clusters of cells form, getting ready to specialize into these organs. The clusters, called primordia, come into existence one at a time. The overall pattern of growth is spiral. Each successive primordium pops into being along a tightly wound generative spiral, and the spacing between each primordium and the next is the golden angle. It turns out that this particular angle leads to efficient packing of primordia, and no other angle performs as well in this respect. However, this efficiency is a consequence of the growth pattern, not a cause. The golden angle spacing is the result of mechanical and chemical influences, which encourage each new primordium to develop in the largest available empty region.

This pattern of development offers evolutionary advantages. For example if the primordia are developing into leaves, then spiral spacing prevents nearby leaves from overshadowing each other. However, there are at least two other patterns—successive primordia may be on diametrically opposite sides of the main shoot, or they may arise in opposite pairs, each pair being set at a right angle to the previous pair. So we have to take evolutionary explanations with a pinch of salt.

Hans Meinhardt has shown that the same activation-inhibition chemical schemes that explain the markings on seashells (*see pp. 124–125*)

RIGHT The seedheads of sunflowers and daisies are created by arranging successive units at angles separated by 137.5 degrees. This angle is special: it ensures that the seeds are tightly packed and evenly spaced (center). If the angle is just a bit too small (left) or too large (right), the seeds won't pack properly.

can also generate these three patterns in the growth of plants, with the golden-angle spiral being the most common. A number of mathematicians and physicists, among them Stephane Douady and Yves Couder, have either mimicked the dynamics of the growth of primordia in physical experiments or simulated them on the computer. Their work has confirmed the importance of the golden angle and its relation to the Fibonacci numbers.

It has been known for centuries that the golden angle is intimately associated with Fibonacci numbers. The simplest way to describe the connection is to form fractions from consecutive Fibonacci numbers— $\frac{3}{5}$, $\frac{5}{8}$, $\frac{8}{13}$, and so on. These fractions of a circle, expressed in degrees, get rapidly closer and closer to 222.5 degrees. This is the golden angle, thinly disguised—the same angle, but measured round the outside instead of the inside. (If you're not following, what I mean is: $222.5° + 137.5° = 360°$.)

Because the golden angle is so closely approximated by Fibonacci fractions, the primordia in a growing plant form geometric structures apparently compatible with Fibonacci numerology which is why so many flowers have a Fibonacci number of petals.

The most dramatic instance of Fibonacci numerology in flowers is the seedhead of daisies, especially large sunflowers. Here the seeds, grown from primordia laid down at successive golden-angle spacings along a generative spiral, line up into more obvious spiral swirls. Typically there will be, say, 34 clockwise swirls and 55 counterclockwise ones—or 55 and 89, or 89 and 144. Consecutive Fibonacci numbers.

The same numerology can also be seen in cauliflowers, which we usually think of as featureless lumps of soft, white tissue. On closer inspection, we find that the lumps are arrayed in beautiful spiral swirls. Sometimes the eye of a mathematician sees things that other eyes miss.

WHIRLPOOLS

While contemplating the elegant spirals of the sunflower, it is impossible not to compare them with other examples of spiral geometry in nature. Spirals arise whenever a moving or a growing system combines rotational movements with radial ones, and it is therefore no surprise to find a variety of spiral structures in nature. Two we've met already—sunflowers and shells. The horns of sheep, goats, and certain types of deer, such as gazelles, often grow in spirals too. But nature's most dramatic spirals mostly lie in the domain of physics, not biology.

One of the great triumphs of early research into statistics was Francis Galton's discovery of anticyclones, published in 1863. He analyzed numerical observations of the weather for December 1861—wind direction, temperature, and pressure. When he plotted the geographical variation of these numbers, over areas roughly the size of the British Isles, he noticed a tendency toward spirals. This was the first serious observational evidence for the idea that the Earth's atmosphere could—and every so often should—form giant whirlpools.

Mathematicians and physicists had already discovered that flowing fluids, such as water, have a tendency to form rotating eddies, in which the flow spirals around a fixed or moving center. The best known example of such eddies is the von Kármán vortex street, which forms behind an obstacle placed in a uniformly flowing stream of fluid. Successive vortices spin off to the left and to the right of the obstacle, rotating in opposite directions, related by glide reflections.

The Earth's atmospheric vortices exhibit a similar tendency, though they are not shed by obstacles and so do not come in pairs. Nonetheless, they form anticyclones, which rotate clockwise in the northern hemisphere and counterclockwise in the southern hemisphere.

The direction is a consequence of the Earth's rotation and the fact that "up" is not at right angles to the Earth's axis—except at the equator. This produces a tendency for the atmosphere to drift in different directions in the northern and southern hemispheres, known as Coriolis forces. And it is these forces that determine the direction of spin.

This fact has given rise to an oft-quoted scientific myth, to the effect that in Australia water spirals down a sink in the opposite way to what happens in Europe and North America. Like many myths, this one has a core of truth. Experiments using huge circular tanks of water, left undisturbed for days so that all movement in the tanks has died down, confirm that each of the Earth's hemispheres has a different preferred direction of spin. However, you won't be able to observe this in the hotel sink, where the effect is swamped by short-term asymmetries such as which tap supplied the water.

Galton discovered anticyclones long before we had high-altitude aircraft or satellites that could photograph them. Nowadays, the spiral tendency of the atmosphere is obvious—especially when the spirals come in their most violent form, a hurricane. In an anticyclone the pressure is highest at the center; in a hurricane—otherwise called a cyclone (and in the Indian Ocean a typhoon)—it is lowest at the center. A hurricane is one of nature's most beautiful disasters, a gigantic swirl of storm clouds fueled by heat and humidity, creating winds up to 125mph (200km/h). A hurricane can flatten tall buildings and devastate entire areas of cities.

On the largest scales of all, nature's most impressive spirals are galaxies. A galaxy is a rotating disk of stars—typically several hundred billion of them—separated into distinct spiral arms like the pattern made by a Catherine wheel firework. Mathematical models of galaxies can reproduce the spiral structure as a consequence of rotation and gravity, but there are problems with how the speed changes at different distances from the center. We still have a lot of work to do before we can claim to understand the dynamics of galaxies.

As well as spiral, galaxies can be elliptical, lens-shaped, or irregular. Usually a spiral galaxy has two arms; often they stream away from a pronounced central bar. At the center of most galaxies, we now think, lies a giant black hole, a gravitational plughole into which matter is being sucked. Is a galaxy the cosmic analog of a bathroom sink? Truth can be stranger than fiction. Watch this space.

FAR LEFT AND LEFT Seen from far enough away, the howling disorder of a hurricane resolves itself into a slowly swirling spiral of warm, moist air, which we then see as a whirlpool of clouds (far left). Intricate flow patterns of water, including turbulent flows, are formed by a series of spinning vortices (left), each one resembling the spiral of a hurricane. Nature builds complex forms out of simple components.

11

TIME

Spiral galaxies, whirlpools, and hurricanes have a lot more symmetry than you'll see in a photograph. In a still snapshot, a spiral shows a definite pattern, but for one-armed spirals only the logarithmic spiral can really be said to possess symmetry. Spirals with two or more arms have rotational symmetries, clicking one arm around to occupy the positions of the other arms. Spiral galaxies are commonly two-armed and are (roughly) symmetric under rotation through 180 degrees. A movie, however, reveals another class of symmetries of these objects, symmetries in time.

The "whirl" in "whirlpool" gives the game away—all of these natural spirals rotate. And to a first approximation they retain the same shape as they rotate. In effect, they rotate rigidly. This implies that they have infinitely many symmetries—not in space alone, not in time alone, but in space-time.

I'll amplify this remark. Let some period of time pass—the spiral rotates through some angle, but does not change shape. Rotate it backward through that angle, and it ends up where it started. So a combination of a shift in time and a spatial rotation leaves the spiral looking unchanged. As long as it's rotating with uniform speed—which is approximately true for these natural forms—the same combination of time-shift and rotation fixes not just a snapshot but the entire space-time trajectory of the spinning spiral. If you were to represent

the motion in a space-time diagram—which stacks successive frames of a movie on top of each other, at right angles to the usual directions of space—then the spinning galaxy would have a perfect helical screw-thread appearance. Its spatio-temporal symmetries are indeed those of the helix.

What sort of symmetries arise when we take time into consideration? Mathematically, time forms a one-dimensional continuum, a line. Lines have two kinds of symmetry—translations and reflections. Translations shift the whole time-line along. They are phase shifts which move the whole system, or its mathematical description, forward or backward in time. Reflections reverse the direction of time— they are like running a movie backward. I've already mentioned that the laws of Newtonian mechanics are symmetric under time translations and also under time-reversal (*see pp. 28–29*), and the same goes for most of the more modern laws of physics.

A steady state has complete temporal symmetry. Let time pass, run time backward— the system wasn't changing anyway, so it still looks exactly the same. An irregular rock that sits unchanged for hundreds of years is displaying a huge amount of temporal symmetry, but this sort of symmetry is boring and we seldom notice it.

In contrast, we get very excited about cycles, particularly those in which the same events repeat over and over again. Periodic cycles have temporal symmetries, too: they look the same if time is translated by any whole number multiple of the period. Spring, summer, fall, winter—

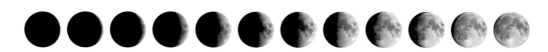

the seasons set in at roughly the same date in any year, be it 1066 or 2001. Translate time by a whole number of years, and nothing untoward would result. Translate time by a fraction of a year, say six months, though, and you'd get winter when it ought to be summer and summer when it ought to be winter.

Some periodic cycles have time-reversal symmetry, others don't. The temperature changes that occur during the cycle of the seasons are approximately symmetric under time-reversal. In forward time, starting in midwinter, the temperature runs from cold to moderate to hot to moderate to cold. Reverse the sequence, and it looks the same. The cycle of the Moon's phases is subtly different. The visible part of the Moon goes from new Moon to crescent to gibbous to full to gibbous to crescent to new, and this sequence looks the same when reversed, but in the northern hemisphere the illuminated side of the Moon is the right-hand side during the first half of the sequence, and the left-hand side during the second half. The true symmetry of the Moon's phases is therefore a combination of space and time—reverse time and reflect space. Many of the time-reversal symmetries of nature are like this.

BELOW Nature often moves in cycles. The phases of the Moon wax and wane in a rhythm that has not changed a great deal since the days when dinosaurs roamed the Earth.

RIGHT The four seasons form a cycle from spring to summer, then fall, then winter.

ANIMAL MOVEMENT

Animals use these space-time symmetries to move.

At slow speeds, horses walk. When they need to go a little faster, they trot. At higher speeds, they canter, and if they need to go full speed they gallop. Patterns of locomotion in animals are known as gaits, and there are dozens of different ones. One thing they all have in common is that when the ground is level and the animal is moving at constant speed, the animal repeats the same leg movements over and over again. Gaits are periodic cycles. Of course animals can also move in nonperiodic ways, for example when negotiating variable terrain, but we don't call these movements gaits.

When a horse walks, it moves its left rear leg, then its left front leg, then its right rear leg, then its right front leg. Each hoof hits the ground at equally spaced intervals, each one quarter of the full gait cycle. In the trot, the hooves hit the ground in pairs—left rear and right front hit the ground together, then right rear and left front. Again, the timings are equally spaced. The canter is much more complicated. First the left rear hits the ground, then half a cycle later the right front; then roughly a quarter of a cycle later the other two hooves hit simultaneously. The gallop is, if anything, more regular than the canter. The left rear leg hits the ground first, followed almost immediately by the right rear. Then, half a cycle after the left rear leg hits, the left front leg hits, followed almost immediately by the right front. (Both canter and gallop occur in mirror-image forms. I'll talk about just one of the two possibilities—the other is the same except that left and right are interchanged.)

If the tiny delay between left and right in the horse's gallop is eliminated so that both rear legs hit the ground, then both front legs hit the ground, we get the bound. Rabbits and squirrels bound. Dogs also bound when they need to move really fast. In the pace (or rack) the two left legs hit the ground together, then the two right legs—the camel and the giraffe typically pace.

At top speed, the cheetah gallops, but its gait is subtly different from the horse's gallop. The horse's left leg always hits the ground before the corresponding right leg does, at both the rear and the front. More precisely, this pattern is called a transverse gallop. The cheetah follows the same pattern with its rear legs, but swaps the order at the front: right then left. This is known as a rotary gallop.

The final quadruped gait that I'll mention is the pronk, in which all four legs hit the ground simultaneously. Young gazelles sometimes pronk, probably to confuse predators. The pronk is extremely symmetric. Swap the legs any way you like, and the timing doesn't change.

The first person to emphasize the symmetries of gaits was the zoologist Milton Hildebrand, who distinguished the symmetric gaits (walk, trot, bound, pace, and pronk) from asymmetric ones (canter and the gallops). To him, the difference was that the canter and the gallops are different

ABOVE AND RIGHT The impressive strides of a sprinting cheetah (above) rely on a simple mathematical pattern of footfalls, which characterizes the rotary gallop. The pattern for a fast horse is subtly different, and is know as a transverse gallop (right). The cheetah's front legs hit the ground in the opposite order to the back legs, while for the horse the order is the same. (The graphs show when a leg is on or off the ground and when a foot first hits the ground.)

from their mirror images, while the other five gaits are the same as their mirror images.

The same? Not quite. If you look at an animal that is bounding in a mirror, there is no significant change, because the left and right sides do the same thing at every instant of time. A pacing animal is subtly different, though. When the pacing animal moves its left legs, its mirror image appears to move its right legs. Both the animal and its image pace, but there is a difference in the timing.

Let's look at another potential symmetry—swapping front and back legs. Again this leaves the pacing animal looking exactly the same, but creates a half-cycle phase shift on the animal which is bounding. What about the trot? Both left-right and front-back interchanges cause a half-cycle phase shift. But if we interchange each leg with its diagonal partner then the phase shift vanishes and the legs hit simultaneously.

The significant symmetries of gaits, then, involve both space (interchange various legs) and time (shift phase). This idea causes us to reassess the walk. If we cyclically permute the legs in the order left rear, left front, right rear, right front, and shift phase by a quarter of a period, we have a spatio-temporal symmetry of the walk. So now we find that we have refined Hildebrand's characterization of symmetry in gaits—each of his five "symmetric" gaits has spatio-temporal symmetry, and their symmetries are different.

RODEO BREAKTHROUGH

Animal gaits are one of the themes of my own research, and therein lies a tale. The space-time symmetries of animal locomotion are very similar to patterns that arise naturally in networks of coupled oscillators—think of lots of pendulums tied to each other with elastic. Each pendulum on its own can oscillate—swing to and fro. The elastic communicates the movement of one pendulum to the others, and it's quite fun to work out just what the combined system actually does.

How did I get into this area? It all began when I was reviewing a book that included some material on legged robots. I mentioned the analogy between legs and coupled oscillators, saying: "Would anyone like to fund an electronic cat?" Within 24 hours I received a phone call from Jim Collins, a young American physiologist visiting Oxford. "I can't fund one, but I know people who can." He took the train to Coventry and we began a collaboration that continues to this day. We still haven't got to build that electronic cat—but I did build an elastic horse for a television program.

The most obvious oscillators in animal movement are the legs, but this is misleading. The source of the patterns in legged locomotion is not the legs as such. The legs reveal what the pattern is, but what really matters is the circuitry in the animal's nervous system that sends signals to the muscles that move the legs. This central pattern generator is a network of coupled neural oscillators, and it probably lies in or near the spinal column (in those animals that have spines).

Jim and I looked at the natural mathematical patterns in four-oscillator networks, finding close analogies with quadruped gaits, and two-oscillator networks, which corresponded neatly to bipeds. Later we extended the work to six-oscillator networks and hexapods (insects). Then my regular collaborator Martin Golubitsky pointed out a technical snag. No four-oscillator network can reproduce the dynamics of quadruped motion in a totally satisfactory way—the analogy has a flaw.

Eventually we worked out that the only sensible design of central pattern generator within this particular mathematical framework must have two oscillators for each leg—one to push and one to pull, so to speak. We still don't know that this idea is right, but at the very least it has led to a lot of interesting mathematics. Moreover, it leads to some specific predictions about real animals, and there is growing evidence to support these predictions.

For instance one prediction is that there should exist a new kind of quadruped gait. In a rhythm of four beats to the bar, this has the following timing:

Beat 1: Both front legs hit the ground.
Beat 2: Both back legs hit the ground.
Beat 3: No legs hit the ground. (They may be on or off the ground, but they don't hit it.)
Beat 4: No legs hit the ground.

We looked, but we couldn't find this pattern described anywhere in the gait literature, but according to our theory it ought to be there. We named it the jump.

It worried us.

Marty is at the University of Houston and when you're in Texas you go to the rodeo. It so happened that the rodeo was in town at the time we were developing the eight-oscillator network, so we went. We were watching the bucking broncos and suddenly we both leaned forward in our seats and started counting on our fingers. First, the horse's front hooves hit the ground. Then, very quickly afterward, the back hooves hit. Then the horse jumped into the air and seemed to hang there…Then it did the same again…and again…until the rider fell off.

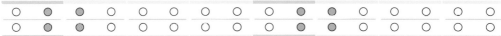

B F B F B F B F

It looked very much like our missing gait. Later we got hold of a video recording of the event, copied it into a computer and looked at it frame by frame—trying to work out exactly when each leg first hit the ground. To within a twentieth of a second, the timing was exactly what we predicted for our jump. We don't claim this evidence is conclusive, but…

ABOVE The violent bucking of a rodeo bronco, doing its best to remove the cowboy from its back, fills in an otherwise missing animal gait from the mathematician's catalog of expected gait patterns. Once this is done, the mathematical catalog exactly matches the patterns used by nature. In the "jump" gait shown in the line diagram above, the animal's feet hit the ground in rapid succession, followed by a pause while the horse is in the air and no feet are touching the ground.

THE SIX-LEGGED INSECT

With four legs, there's room for an awful lot of different gait patterns. With six legs, the number found in insects, there ought to be room for a whole lot more. The same mathematical theories apply to four legs, six legs, or for that matter, to two or a hundred. The gaits observed in real insects fit neatly into the resulting catalog of the patterns formulated by theory. The favorite experimental animal in this area is the cockroach. At slow speeds, the cockroach adopts a six-legged version of the quadruped walk. Its legs move in the order: back left, middle left, front left, back right, middle right, front right. Each step is equally spaced in time, like the footfalls of the horse or the elephant. This pattern is called the metachronal gait.

When a cockroach needs to move fast it employs the tripod gait, a sort of six-legged trot. The legs group together to form two sets of three, and all legs in each given set move together. One set consists of the front and back left legs as well as the middle right; the other set consists of the rest: front and back right legs and the middle left. These overlapping tripods of legs move alternately, at equally spaced intervals. The tripod formation has major advantages—the same ones that lead human photographers to rest their cameras on tripods. The tripod is stable and can rest on uneven terrain.

Beyond insects come spiders, with eight legs; crayfish, with ten; centipedes and millipedes. Centipedes are especially interesting, and their greater number of legs makes certain regularities of gaits more obvious to the observer. We've seen that when a centipede moves, rhythmic waves of activity ripple along the legs on each side of its body (pp. 40–41). These waves move from the back to the front, and the wave on the left is antisynchronous with the wave on the right—just as the legs in the human walk are antisynchronous. Several complete waves can fit into the length of the centipede's body.

Like the horse, the cockroach, and the human, centipedes too have several "gears." At low speeds, three or more complete waves fit along the length of the body. As the speed increases, the number of waves traveling down each side

becomes smaller, the body wiggles sinuously from side to side, and at any given moment the weight is supported by fewer legs. In wild centipedes traveling at high speeds, there are times when only 3 legs are in contact with the ground—out of a total, in this case, of 40!

All of the legged gaits that we have described have common features, despite their apparent diversity. They all consist of a pair of traveling waves of leg movement. One wave travels along the left side of the animal; the other wave travels along the right side. Usually the waves start from the back and move toward the front. There are two main types of wave pair. Either the waves on both sides are synchronous or they are out of phase with each other. Mathematical models of central pattern generators show that all of the apparent diversity of animal gaits is basically an elaboration on this basic common structure.

Where does this common structure come from? I reckon it's evolution. The early ancestors of four-legged mammals and six-legged insects were arthropods, creatures with many segments. Each segment bore two legs, one on either side; all segments were pretty much identical except

at the head and tail. Along with the legs, the segments also bore the neural circuitry required to control them. The mathematical model of a central pattern generator mimics the symmetry of an arthropod; its natural patterns of oscillation are traveling waves, with the two sides either in phase or out of phase.

As the arthropods evolved, they dropped segments (and their legs), fused segments together or modified segments for special tasks. An insect's head is 6 fused segments, its thorax is 3 fused segments, each with a pair of legs, and its abdomen is from 8 to 11 segments. An elephant can be seen as the remaining two segments of a distant arthropod ancestor—and it may well carry with it relics of that ancestor's gait-control circuits. Recently the American biologist Randy Bennett and colleagues discovered that if two genes, called Ultrabithorax and Abdominal-A, in the flour beetle larva are switched off, the creature develops 22 segments. So that old ancestral architecture still lurks within modern animals' genes, but usually it is suppressed.

LEFT AND ABOVE A centipede (left) may not have a hundred legs, but it still has an awful lot. How does it walk? Waves of movement travel along each side of the insect, and the positions of these waves alternate left and right (above). A millipede moves in a similar way, but now the two waves are synchronized.

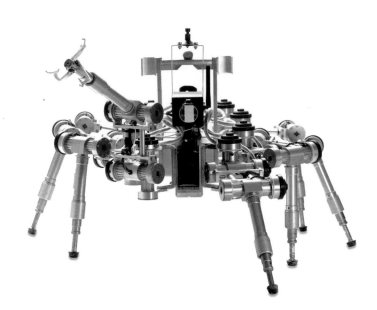

ROBOTS WITH LEGS

Why should we try to understand gaits, anyway? Will it lead to a better breed of horse? No, but it may well lead to a better breed of robot.

Today, nearly every car in the world is assembled with the help of robots, but they don't look at all human (unlike the robots featured in the science fiction of Isaac Asimov). Industrial robots are mostly fixed; when robots need to move, they mostly run on wheels. However, research into legged robots is growing fast, and roboticists are finding it worthwhile to take some tips from nature. By understanding how legged animals move, they can build better legged machines.

What's the point of making a robot with legs? There are many tasks that involve moving around in terrain that is unsuitable for wheels and dangerous to humans. Army firing ranges the world over are littered with unexploded ordnance. The ground is usually very poor—rocky, covered in thorny scrub—because you don't waste good land by putting a firing range on it. The task of finding unexploded shells and making them safe is too dangerous for a human being, a cheap robot would be far better. Making minefields safe is another potential application, basically the same one in slightly different circumstances. Another potential application is the decommissioning of nuclear power stations. This will have to be done remotely by machines because exposure to high levels of radioactivity is fatal for people. The terrain inside a nuclear power station, especially one in the process of being dismantled, is unsuitable for wheels.

A more exciting application is planetary exploration. There are no roads on Mars or Venus—yet. All robot explorers used to date have had wheels, notably the six-wheeled Sojourner on Mars, but legs would be better if they could be made reliable enough. And therein lies the problem. With the current state of knowledge, legged robots tend to break down too frequently, and keeping them upright is also tricky.

Around the world, especially in the United States and Japan, roboticists are working to design and build practical legged robots.

One is Mark Tilden at Los Alamos National Laboratory in New Mexico. He specializes in small, self-contained four-legged robots whose insectlike movement is driven by solar power. He has also built a legged robot the size of a small dog that hunts unexploded mines and shells. Ideally, it detects them without exploding them and reports their location. Sometimes, though, it steps on one and loses a leg. No problem, it's a robot—and Tilden's robots are robust. His robots are designed so they can get around with three legs. They can get around with two legs. They can even claw their way along with one leg. So each explosion costs one-quarter of a robot.

At Case Western Reserve University in Cleveland, Ohio, there are three robots, rather imaginatively named Robot One, Robot Two, and Robot Three. They were developed through a collaboration of Randall Beer, a computer scientist, Roy Ritzman, and Hillel Chiel, who are biologists. These robots are six-legged, and the system that controls their movement is modeled on the nervous system of an insect.

Understanding insect movement has greatly improved the robots' performance. But the flow of ideas goes the other way, too. The ability to make precise measurements of the forces acting on the legs of the robots when they move has led in turn to a better understanding of how insect legs work.

Another American, Joseph Ayers at Northeastern University has built a robot lobster with eight legs. The aim of this is entirely practical. He sees it as a route to building underwater remote-sensing vehicles. Similarly, Michael Dickinson at Berkeley is working on building a robot fly. And Michael Tiantafyllou at MIT has built a robot tuna in order to study how fish make use of underwater vortices to enable their propulsion.

Tilden has a more ambitious vision. He sees tiny, autonomous, solar-powered robots as the future of space exploration. Make them in their thousands, equip them with communicators to relay messages to a higher-powered base station, and set them loose on Mars to report whatever they find.

BELOW AND RIGHT At
supersonic speeds
Concorde (below) used
vortices to generate
lift, which trails across
the wings and behind

the aircraft. However,
according to classical
aerodynamics, neither
Concorde nor a bee
(right) ought to be able
to fly. Modern theories of

flight are more subtle and
explain how they do it.
The bee's rapidly moving
wings create a vortex that
spirals along the leading edge
of the wing, generating lift.

FLIGHT OF THE BUMBLEBEE

We've looked at how creatures—animate and
inanimate—move on legs, but what about those
that have wings?

Flight literally provided a new dimension to
evolution—up. So advantageous was this new
dimension that it was discovered many times—
by insects, by bats, by fish that glide for short
distances on aerodynamic fins, by other gliders
such as the flying fox and assorted squirrels,
frogs, lizards, and even snakes. And—
of course—by birds, the sport's most elegant
and accomplished exponents. When evolution
first discovered flight, organisms tried all sorts
of strategies for getting off the ground and doing
something useful once they were up there.
Among the more extreme examples, later
discarded because they couldn't survive the
competition, were dragonflies with 20-in (50-cm)
wingspans, found as fossils from the Permian

period and probably dating back to the
Carboniferous period.

Birds did better. They seem to be some kind
of offshoot of the dinosaurs (a theory that
was considered insane 20 years ago and is
conventional wisdom today), but their family
tree branched off many tens of millions of years
before a giant meteorite smashed into the
Yucatán coast and put an end to the dinosaur
dynasty. Recently it was suggested that they
branched off earlier still, from reptiles, but that's
hotly disputed. The main piece of gadgetry that
makes birds feasible is not so much the wing—
basically just a modified leg—as the feather.
Feathers are lightweight and strong, just the
thing for effective flight.

Most of what goes on when a creature flies is
invisible to the eye, because air is transparent.
A bird in flight spins off a regular pattern of
vortices, swirls of air spinning off from the
trailing edges of its wings. The bird exploits

these vortices to gain lift. Only recently have human engineers understood this particular trick, but the birds have known it for a hundred million years.

In fact, birds use at least two "gaits"—distinct patterns of vortices. If you look at the wings, all you see is a rhythmic bilaterally symmetric sequence of movements—the avian analog of a quadruped bound (*see pp. 132–133*). But if you could see the air (or if you use smoke or other trickery to see how the air moves) you find that some birds create a wake of disconnected ring-shaped vortices, like smoke rings, while others produce a connected wake in which successive vortices merge into each other.

It has taken humans a long time to work all this out. Not so long ago, it was commonly stated that a bee is an aerodynamic impossibility. Its wing is the wrong shape and the surface too tiny to generate enough lift to keep the insect in the air. Bees, undeterred, failed to fall out of the sky at the revelation. However bees manage to fly, they are not tiny fixed-wing aircraft, and they have no reason to obey those parts of aerodynamics that humans have managed to puzzle out.

Now, however, we know how bees and other insects pull off their counterintuitive feat. Their wings move upward until they almost touch. When they beat downward, the sharp front edge creates a leading edge vortex. For reasons we still don't fully understand, this vortex remains "stuck" to the top of the wing, generating lift, and spirals along it until it is shed at the wingtip. This method requires small wings that beat very rapidly, which is why bees buzz. And it's why the flying pig will remain a metaphor for the incredible.

BIZARRE LOCOMOTION

People who keep pets fall into two main kinds. Some stick to conventional animals—cats, dogs, budgies. Often they own just one kind. Others go for the exotic—lizards, snakes, or tarantulas. Like as not, if they own one such beast they own a dozen, all different, and all weird. Why? Because it is exoticism itself, rather than any particular instance, that these people value. They like difference for its own sake—and that leads inevitably to diversity.

Collectors of the bizarre in ways of animal movement can find much to entertain them. There is, for instance, the jumping maggot, the larva of the fruit fly. *Ceratitis capitata* has the ability to bend itself into a tight U, clasp the ends together while generating considerable elastic tension, and then suddenly let go…grubs away!

It must give predators quite a shock.

Then there's the sidewinder. Most snakes slither, in a series of sinuous S shapes. Not the sidewinder, which inhabits the desert lands of Mexico and the southwest United States. This snake has other ideas. It flips its body into the air in a series of coils, in effect "rolling" sideways like an angular helical spring rolling across a flat surface. It leaves distinctive traces of its passage, a series of roughly parallel diagonal lines where its body has made contact with the hot sand. Why does it adopt this strange movement? Possibly, to reduce contact with that hot sand.

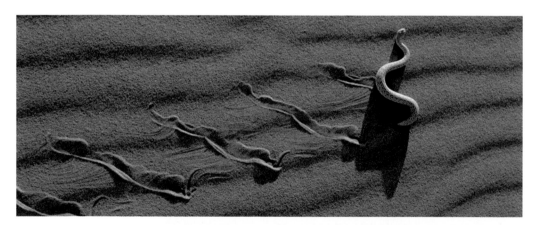

ABOVE AND RIGHT Many different ways to move can be observed in the animal kingdom. Sometimes creatures with legs behave as if they have wheels: for example the wheel spider, which turns its legs into spokes and rolls (right). The sidewinder snake (above) makes as little contact with the hot desert sand as possible by flipping itself sideways and forward, rather like a corkscrew.

The moving machinery of humankind is based almost exclusively on the wheel. There are few wheels in the animal kingdom, probably because there are also few roads. The most important technological advance in making wheels is to build flat roads for them to roll on. One creature that does use a wheel, albeit a spoked one, is the wheel spider. Usually this tiny desert creature scuttles about on eight legs like any self-respecting arachnid. When it is in a hurry, though—attempting to escape a predator, for instance—it tips on to its side, bends its knees to make a broad "rim," and rolls down the sand dunes with impressive speed. The caterpillar of the mother-of-pearl moth also coils up in a wheel, rolling backward at 40 times its normal speed when it needs to make a quick exit, but it manages only four or five complete revolutions. Perhaps the most surprising of the organic wheels is a tiny molecular motor used by some bacteria to make their tails spin. *Escherichia coli* bacteria have flagella, long whiplike protrusions which appear to wave from side to side.

In 1974 the American microbiologist Howard Berg realized that *E. coli* flagella are rigid. They are helices and they rotate. Seen from the side, a spinning helix looks like a wavy sine curve, and a waving tail is so plausible that observers saw what they expected to see. The full story is even stranger. At the base of the helix is a diminutive circular "rotor," a molecular bearing into which the end of the helix plugs, permitting it to rotate. So a bacterium is rather like a tiny ship, with its own screw propeller at the stern. The mathematical formalism used to understand the motion of microorganisms in water is very different from that used for larger objects, requiring a substantial rethink of conventional fluid dynamics. The organisms are so tiny that inertial effects play virtually no role—what matters are viscosity and Brownian motion.

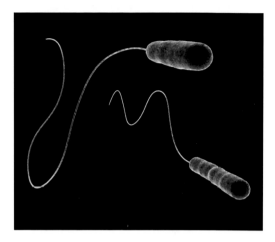

ABOVE Bacteria are so small that to them, water is like a thick syrup. As a result, they have been forced to evolve their own special ways of moving. One solution they have evolved is a helical "screw," much like that on a ship, which is rotated by a tiny molecular motor.

Viscosity is the "stickiness" of water, which is very great if you are a bacterium, and Brownian motion is random buffeting by molecules as they move to and fro. For larger objects, the effect of viscosity is less—though still important—and that of Brownian motion is negligible. Because of these effects, *E. coli* bacteria cannot go hunting for nutrients, but must wait for them to be randomly nudged in its direction. An individual bacterium can stop in an extremely short space—when its motor stops turning, the bacterium comes to a halt within a distance smaller than the diameter of a hydrogen atom. Or, rather, it would come to a halt, but the buffeting molecules surrounding it jolt it this way and that through its own width every second. Current attempts to understand how these tiny organisms are able to move themselves around in such a medium involve dynamical concepts that did not exist two decades ago.

FORWARD TO THE PAST

Periodic cycles have a temporal symmetry—shift time by any whole number multiple of the period, and the behavior is totally unchanged. There is one other type of temporal symmetry with deep physical and philosophical significance—time-reversal—"reflection" in a "time-mirror."

What happens if we run time backward? A world running in reverse seems bizarre. Eggs "unboil." Fragments of a broken plate skid together on the floor and reassemble themselves. An adult human shrinks in size, turns into a child, then a baby, and finally is absorbed by its mother.

Unthinkable? In principle, these things could happen in our universe. According to the laws of physics, the universe reflects realistically in time-mirrors—the laws of nature are the same whether you run time forward or backward. An egg can unboil, a plate can reassemble, and an adult can turn into a child. These events are perfectly consistent with all of the fundamental rules by which the universe operates.

However, we perceive the universe as having a definite arrow of time, one in which eggs boil, plates smash, and children get born and grow up. Our consciousness slides along that arrow at a rate of one second per second. It seems difficult to reconcile our own one-way universe with its two-way, reversible laws. So is there a genuine forward/backward time symmetry in the whole universe, which we are somehow failing to observe?

Or has the universe itself chosen a definite time arrow?

Part of the answer is that the laws of nature can be time-reversible even though our universe has a definite arrow of time. What reversibility of the laws implies is that another universe could have the opposite arrow of time from ours while obeying exactly the same laws. If the behavior of our universe were time symmetric in all possible time-mirrors, then nothing would ever happen and time would have no meaning.

Why, then, does our universe choose to smash plates but not unsmash them? Well, maybe we just live in that one, and it could have been the other way around, as in Martin Amis's *Time's Arrow* and the earlier *Counter Clock World* by Philip K. Dick. Or maybe it's just us. If we could run our minds in reverse, then all events would seem to us to run backward. If this is true must we assume that it is our consciousness that points time's arrow in one particular direction?

If we give serious consideration to setting up some time-reversed scenario in this universe—such as making a plate unsmash—we soon realize that there is a crucial difference between a plate being dropped on the floor and smashing into fragments, and the fragments of a potential future plate hurtling together and assembling. A smash has a completely local cause. You drop the plate—bang. It's all triggered by a single action and distant parts of the universe do not have to conspire in implausible synchronicity to make a plate smash. However, if you wanted to arrange for a plate to reassemble itself, you would have to take all of the separate pieces—which just happen to fit together—and provide each one with the right impulse to set it on a perfect collision course with all the other pieces. At the same time you would have to create distant vibrational waves in the floor so that they all arrive together at the future plate at the exact instant that it reassembles, hurling it into the air to be caught by a waiting hand. So the chain of causality for an unsmashing plate is not local. It involves synchronized events at distant locations.

This gives a way to distinguish the two directions of the arrow of time. In the human universe, most causality is local, while nonlocal

ABOVE Physicists have long been puzzled by a strange symmetry of the laws of nature: time reflection. If time runs backward, the universe still obeys the same laws. A universe that is expanding, as a result of the Big Bang, becomes a contracting universe when time runs backward, ending in the Big Crunch.

LEFT If all the pieces of a broken lighbulb could be hurled together at exactly the right speed and in precisely the right direction, then the lighbulb would "unsmash" itself and once more be whole. But even the vibrations of individual atoms would have to be controlled to achieve this.

causality is much rarer. Our own consciousness—or just our ability to affect our environment—functions in terms of local causality. We notice things like plates, and can pick them up, but we don't notice that scattered fragments might just come together to unsmash on a suitably vibrating floor, and we can't make them do that anyway. So the arrow of time may arise because we are local-causality beings, obliged to experience the broken symmetry of the universe in one—and only one—of the two possible directions.

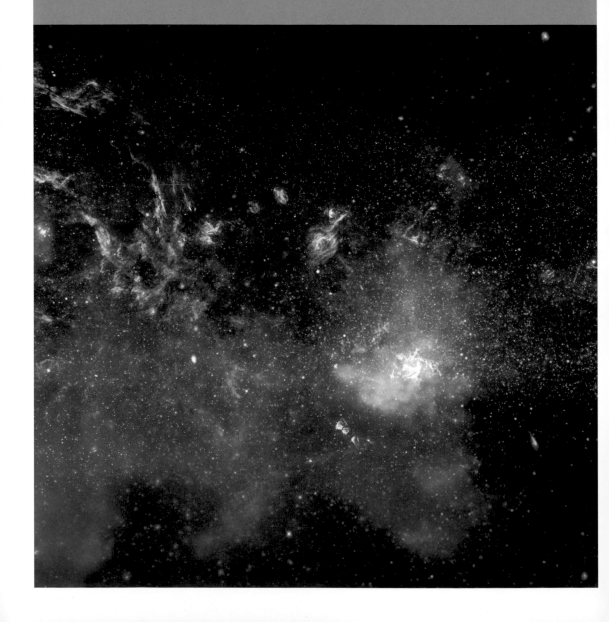

PART THREE

SIMPLICITY
& COMPLEXITY

12

COMPLEXITY & CATASTROPHE

Our search for the sources of nature's patterns is uncovering philosophical depths. The old idea was straightforward—the universe may look complex, but everything is following simple mathematical laws. The regularities of nature that we call patterns are evidence for, and direct consequences of, the simplicity of these laws. However, it's beginning to look as if the link between laws and patterns—between cause and effect—isn't that straightforward.

If it was that straightforward, then simple laws operating in simple circumstances would always lead to simple patterns, while complex laws operating in complex circumstances would always lead to complex patterns. Indeed, there would be a principle of "conservation of complexity," whereby the complexity or simplicity of a cause would necessarily be passed on unchanged to the resulting effect.

This no longer looks correct, but it's taken time to find out because we seem to be predisposed toward such a principle. We ask where simplicity and complexity come from, as if they start out elsewhere and are transferred to whatever we're talking about. We expect the source of a pattern to be some other kind of pattern, but on a deeper level. We're happy that a simple law of gravitation creates elliptical orbits, but we'd be unhappy if a simple law of gravitation created irregular orbits, or if a complex law of gravitation created elliptical ones. This is one of the main things that puzzles us about time-reversed universes—we can't see where the nice unsmashed plate comes from in a backward-time scenario (*see pp. 144–145*).

The needs of science and technology, and the interests of scientists and mathematicians,

are changing fast. Every day brings new kinds of problem, new ideas, new questions. Researchers working on today's scientific frontiers have been forced to recognize that simple causes can often give rise to complex effects, and complex causes can often give rise to simple effects. But—to maintain an air of paradox—in an incredibly complicated way.

One of the new mathematical systems that has driven this point home is a kind of mathematical computer game known as a cellular automaton. The game starts with a grid containing colored squares. Each "move" in the game changes the colors of the squares according to fixed rules—for example: "If a red square is surrounded by three green squares and five yellow ones, turn it blue."

It is easy to set up such a system in a computer, run it, and see what happens. It is far harder (in fact often still impossible) to give a good explanation of the results. For example a cellular automaton known as the "Game of Life," invented by John Horton Conway, employs only two colors, black and white, and three short rules. However, "Life" can do anything that a computer can do—calculate the digits of π, list the prime numbers one by one, search the text of this book for the word "Conway." It carries out such computations extremely slowly, but speed is not the issue. Thanks to deep mathematical discoveries of Kurt Gödel and Alan Turing about insoluble mathematical problems, one consequence of this is that even though we know the rules for "Life," we can't predict what any given configuration of black and white squares will do. Given some initial configuration, will the rules eventually remove all black squares and leave a plain white "dead" grid? In general, there is no way to know. All mathematics can tell us is that it is a question beyond the power of mathematics to answer. That's about as complex an effect from a simple cause as you can imagine.

Cellular automata can be thought of as dynamical systems with finitely states (the colors of the cells) with added "spatial" structure. As such they are widely used to model ecosystems. Perhaps a red cell indicates a hungry fox, a blue one indicates a well-fed fox, a gray one indicates a rabbit, a green one indicates a plant. The rules are designed to reflect ecological realities. A hungry fox square will head toward the nearest gray rabbit square, consume it and turn into a blue well-fed fox square. Gray squares will browse on green ones, and so on. Each square can have a life and death cycle and a number of different states—alive, dead, hungry, fed, ready to breed. Tune the rules, and you get surprisingly lifelike results. Using such techniques, the computer can provide genuine insights into what a complex ecology—a grouse moor, a rain forest, a coral reef—is doing.

BIFURCATION & CATASTROPHE

One treasured intuition bites the dust— complexity can emerge from simple rules, it doesn't have to be built into the rules themselves. Conversely, systems that in fine detail can seem very complicated, such as a coral reef or a rain forest, may exhibit simple patterns of behavior on a large scale. A currently fashionable and also rather important new area of mathematics, the theory of complex adaptive

systems, is attempting to understand such emergent phenomena.

Another treasured intuition of ours is that small changes in a cause should produce small changes in its effect. This intuition occurs in biology, the formation of new species, for instance. Mutations are gradual, but a new species is a dramatic change. So evolutionary biologists look for dramatic causes for new species, like a huge flood or the eruption of a new mountain range. Like the intuition about complexity, the belief that gradual causes have gradual effects is false. Well, not completely so— usually small changes in a cause produce small changes in its effect; but sometimes a single small change in a cause can produce a big change in its effect. And although the latter instances are rare, they can still be unavoidable.

We knew this long ago. We just didn't realize there was a mathematical consequence. Take, for example, the proverb "the last straw breaks the camel's back." If you slowly pile more and more straws on to the camel, you are setting up a slowly changing cause. For a long time, the effect changes slowly too—the camel seems not to notice the extra burden. But as the straws accumulate, the animal starts to suffer. Its back sags, its legs begin to shake…Suddenly, the addition of one final straw brings the beast crashing to the ground. Effects like this, in which a system that is in some critical state can

LEFT AND BELOW Mathematical catastrophes—sudden changes with gradual causes—may or may not correspond to real disasters. Sometimes they do. The Tacoma Narrows bridge collapsed when increasingly haevy winds caused it to flex, and eventually break (left). Sometimes they don't. The "caustic" in a coffee cup is a sudden change in brightness, caused by the geometry of light reflected by the cup (below). The underlying mathematics is the same, but the interpretation is very different.

suddenly change to a very different state as a consequence of some small external event, are called bifurcations or catastrophes. Back in the 1960s a branch of mathematics known as Catastrophe Theory came into being, an attempt to put a degree of order into this whole area. Its creators René Thom and Christopher Zeeman, along with many other mathematicians, physicists, and engineers, came up with a short list of "typical" geometric shapes associated with sudden changes of this kind and attempted to apply them to many other areas of science.

One area was animal behavior. Zeeman provided qualitative evidence that suggested aggression in dogs could be modeled by a catastrophe. Under the influences of fear and rage, the dog's mood could suddenly snap from timidity to aggression, and back again. Nobody did any experimental work on dogs, but it turned out that aggression in territorial fish fitted the model beautifully.

A rigorous analog of the camel's back occurs in engineering, in connection with how a bridge buckles or collapses if it is subjected to too great a load. One of Thom's archetypal catastrophes is indeed involved here. A particularly pretty application is to optical caustics. These are bright curves created when light is brought to a focus. You can see a caustic if you look at the top of a full cup of coffee on a sunny day. You will see two bright curves, coming together at a sharp point.

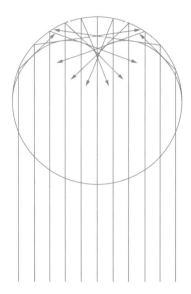

The curves are caused by light bouncing off the shiny circular rim of the cup, and the detailed geometry is precisely that of one of Thom's catastrophes. A caustic with a rather different shape concentrates the sunlight bouncing back from a raindrop into a bright cone, which, you'll recall (*see pp. 66–67*), is what causes a rainbow to be circular.

The word "catastrophe" has gone out of vogue—with its connotations of disaster, it was never a very good choice. The more neutral term "bifurcation" is now the accepted term. If the state of a system changes dramatically as the result of

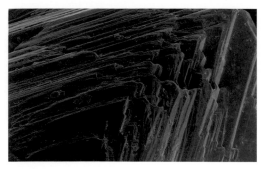

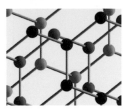

small external changes, scientists say that a bifurcation has taken place. There is an extensive and powerful theory of bifurcations, which allows us to understand these sudden changes in a wide variety of dynamical systems. In fact, the formation of snow is a bifurcation in the state of a system of water molecules, as we shall now see.

PHASE TRANSITIONS

The formation of snowflakes involves an important kind of bifurcation—freezing. As the temperature of water drops below the freezing point 32°F (0°C), liquid turns to solid. A small change in water temperature produces a big qualitative change in the molecular structure and physical properties of water. There is another big change of this kind—boiling. If the temperature of water increases past the boiling point 212°F (100°C), the water turns to steam. Major changes in the physical state of matter, such as liquid to solid or liquid to gas, are called phase transitions. They are bifurcations, but of a highly complex kind, because they involve the collective bulk behavior of huge numbers of atoms.

ABOVE AND BELOW Phase transitions occur when matter changes its state in a dramatic way. Although also made from carbon, diamond (above) is harder than graphite (below), because its crystal structure is different. Graphite is a soft form of carbon that has a crystal structure made from soft honeycombs. It can in principle undergo a phase transition to diamond, but only if huge pressures are used to overcome the large energy barrier between the two states.

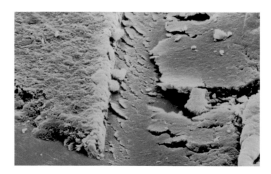

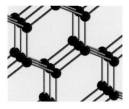

Crystallization is a phase transition. It can occur when a molten solid cools or when a solid is dissolved in a liquid which subsequently evaporates. A given substance may have several solid phases, depending on the symmetry (or otherwise) of its atoms. Carbon, for instance, can crystallize into either graphite (soft, black, looks like dirt and is worth much the same) or diamond (hard, transparent, and ridiculously overvalued). The difference is not in the atoms—both graphite and diamond are pure carbon, but it is how these carbon atoms are arranged. In diamond, each carbon atom is bonded strongly to four others in a regular tetrahedral shape. The atoms pack closely together in what is essentially a cubic lattice and, as a result, they form a very stable structure. In graphite, each carbon atom is bonded strongly to three others in a flat hexagonal lattice, and only weakly to a fourth. The atoms form parallel layers that can easily slide over each other and therefore form a less stable structure. This is what makes graphite soft. In much the same way, ice is crystallized water.

There are at least 16 different crystalline structures whose form depends on pressure and temperature. The normal form of ice is ice I, and the others are ice II up to ice XVI. (In Kurt Vonnegut's novel *Cat's Cradle*, the end of the world is caused by the oceans turning into "ice-nine" at room temperature after a tiny lump of the stuff is dropped in by accident. Fortunately, real ice IX wouldn't have that effect.)

Magnetism is also a phase transition. We all remember playing with magnets at school and discovering how peculiar they are, surrounded by some invisible pattern of forces that makes compass needles point north and aligns iron filings into strangely beautiful patterns. We could actually feel one magnetic north pole repelling another north pole, but attracting a south pole. What we saw and felt were effects of a magnetic field, as real as anything else in the universe, but mysterious to us because it is not directly accessible to our senses.

Bulk magnetic fields arise when the tiny magnetic fields of electrons line up and reinforce each other. Heat causes the atoms to vibrate, destroying the alignment. In certain materials, known as ferromagnets because one of them is iron, there is a distinct phase transition from "magnetic" to "nonmagnetic" as the temperature increases past a particular value, the Curie point (about 1,400°F/750°C for iron).

Why does matter form distinct phases?

Physicists have gained a good understanding of phase transitions through the study of special, simplified mathematical models. The best known of these is the Ising model (named after Gustaf Ising, the Swedish physicist who came up with the analysis of the model in 1925). It employs a square lattice in the flat plane, each vertex of which can be in one of two states—

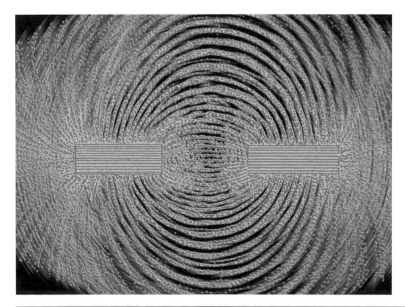

LEFT Magnetization is another kind of phase transition, and the magnetic properties of materials can be destroyed by heating them.

"up" or "down." Physically these options correspond to the directions of the spin of electrons. There are interactions between neighboring vertices, so each electron's spin is affected by the spins of its neighbors. It turns out that at a critical temperature, which can be calculated exactly, the pattern of spins changes abruptly—a bifurcation. The Ising model shows that phase transitions are associated with a change in the symmetry of the material—but it is a symmetry of a very strange kind, statistical symmetries of properties-on-average rather than of individual components.

In ice and carbon, the fine details of phase transitions are considerably more complicated than those of the Ising model, but again we can understand the phase transitions as changes in the symmetry of large collections of molecules.

Different symmetry types of crystal also have different energies, which change under the influence of pressure and temperature. The important changes are the bifurcations, where major qualitative changes suddenly occur. It is these that separate the different phases.

SYMMETRY BREAKING

We now come to the heart of the entire story.

The changes in symmetry that occur in solid-to-solid phase transitions open up a question that is close to the heart of all modern theories of pattern formation. Like complexity and catastrophe, it tickles our intuition and causes us to make major revisions to some cherished, but seldom explicit, beliefs.

In 1894 the physicist Pierre Curie, codiscoverer of radium with his wife Marie, stated a fundamental physical principle. Symmetric causes, said Curie, will necessarily give rise to equally symmetric effects. Conversely, if you see an asymmetric phenomenon, then you should look for an equally asymmetric cause.

I'll take two real-world examples where Curie's principle should apply. It applies to both of them, but in one case it leads in a very unhelpful direction. In this case Curie's principle is technically correct, but misleading—and its opposite gives a better picture of what happens.

The first example is ripples on a pond. In a mathematician's reflex idealization, the pond is an infinite, uniformly thick layer of water, symmetric under all rigid motions of the plane—translations, rotations, reflections. Now, we toss a pebble into it—a mathematician's point pebble. It hits and causes expanding rings of ripples. The cause—the pebble—is not symmetric under all rigid motions of the plane. Neither is the effect. In fact, because the cause picks out one point in the plane as being special—the place where the pebble hits is different from all others—the only symmetries of the cause are rotations about that special point plus reflections in mirrors that pass through that special point. Lo and behold, the effect—circular rings—has precisely those symmetries. That's why we get rings. Score one to Curie.

In contrast, imagine taking a ping pong ball to the bottom of the ocean. Again, let's think about a mathematician's idealization. The ball becomes a perfect spherical shell of elastic material. Way down in the ocean, the ball is subjected to enormous compressive forces. The ball is so tiny in comparison to the ocean depths that we can assume the forces that are trying to squash it flat are the same everywhere, directed precisely toward the center of the ball.

What happens? The causes are spherically symmetric, so according to Curie's principle, the ball should remain spherical. Does it shrink, forming a smaller ball? I don't believe that and neither do you. In practice, the ball will crumple. The mathematics of crumpling is fascinating, and in this case it shows that when the ball first begins to crumple, its surface acquires a system

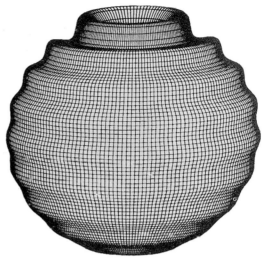

ABOVE AND RIGHT Pierre Curie claimed that symmetric causes always produce equally symmetric effects... but sometimes they don't. If a pebble is dropped into a pond, the causes are circularly symmetric, and so are the effects—a spreading sequence of ripples that move outward across the pond (above). If a hollow sphere is uniformly compressed, then the causes are spherically symmetric, but the effects are not: the sphere buckles (right). Remarkably though, the buckling pattern is circularly symmetric. This is an example of symmetric breaking.

of ripples. The ripples have some symmetry, but not spherical. Instead, they have rotational symmetry about some axis passing through the middle of the ball.

Nonetheless, Curie is technically correct, because the crumpling has to be triggered by something. The trigger is some imperfection in the material of the ball—maybe it's slightly thinner somewhere, or slightly thicker, or one part is a little weaker or stronger than the rest. Another triumph for Curie, then? Yes, but. First, the imperfection can be so tiny that you can't detect it. Second, whatever the imperfection, the initial crumpling pattern has rotational symmetry. As the crumpling proceeds, it quickly loses that symmetry and you get a crushed mess, but you can stop that from happening by putting a slightly smaller solid ball inside.

Curie's principle leads us to seek a rotationally symmetric cause for that crumpled state.

However, no such cause need exist. Within the accuracy of observations, Curie's principle can explain either a spherical ball or a crumpled, totally asymmetric one. What it has serious trouble explaining, though, is the actual pattern that occurs here, with its counterintuitive rotational symmetry.

This phenomenon—of effects having less symmetry than their causes—is known as symmetry-breaking. Its explanation involves another element—stability. Deep in the depths of the ocean, the spherical state of the ball is unstable—any disturbance will destroy it. But the rotationally symmetric crumpling pattern of the ball is stable, at least for a range of forces after the crumpling point. So Curie's principle needs to be modified: symmetric causes produce equally symmetric effects, unless those effects are unstable, in which case the symmetry will break.

WHERE DOES THE SYMMETRY GO?

Symmetry-breaking seems very curious, and it runs counter to naive intuitions about symmetry.

So how does it work?

Where does the symmetry go when it breaks?

Well, it doesn't necessarily go anywhere. Complexity is not conserved and continuity is not conserved, so why should symmetry be conserved? But, in fact, it is—in a subtle way.

When that ping pong ball first crumples, it settles into a state that has rotational symmetry about some axis. The question is, which axis? The mathematics tells us that in principle it could be any axis. As well as crumpling about an axis that runs from top to bottom, say, the ball could equally well crumple about an axis that runs from left to right, or front to back, or any other line that runs through the center of the sphere. The reason for any possible axis is that the equations for the crumpling sphere faithfully reflect its spherical symmetry—they are spherically symmetric equations.

What does this mean? It means that if you take some solution of these equations and rotate it in space, you get another equally acceptable solution. The "mistake" in the over-literal interpretation of Curie's principle is to assume that this rotated solution must be the same as the original one. If you knew that there was only one solution—which is a very common assumption in classical dynamics—then Curie would be right. But dynamical equations can have lots of solutions, and that's what's happening here.

In short, specific solutions may break the symmetry of the original system, but the sum total of all possible solutions remains symmetric. The symmetry gets "spread out" over a lot of solutions. And the trigger for this kind of spreading is the onset of instability in the symmetric solution.

Let's interpret these ideas for sand dunes. The basic system is a flat, infinite plane of sand, over which a uniform wind is blowing. The desert is symmetric under all rigid motions, but the wind direction rules out rotational symmetries (because you can't rotate the wind) and most reflectional symmetries (all mirrors that are not parallel to the wind). The unpatterned desert is a state with the same symmetry as the system just described. However, this state can become

LEFT Symmetry breaking also explains why flat sand in a flat desert responds to uniform winds by making nonuniform patterns. In fact, it can make a wide variety of patterns, depending on the conditions.

unstable. Instead of tiny random humps flattening out, they can grow.

What happens then?

According to the mathematics of symmetry-breaking, the dune sand will shift to whatever new state is stable. Typically, this state will possess quite a lot of symmetry—a good rule of thumb is that systems break symmetry reluctantly, hanging on to as much as they can for as long as they can. (There are exceptions, but these are few and far between.) Suppose, for example, that the translational symmetry breaks. Then the state is no longer uniform at all points. However, the desert could retain some translational symmetries, say through a fixed distance along the wind direction, and hence through all *integer* multiples of that distance. And it could retain all translations at right angles to the wind, say. We've just described the symmetries of transverse dunes— regularly spaced, parallel ridges of sand at right angles to the wind. Or maybe it doesn't retain translations at right angles to the wind, but in some other direction. Now we've just described linear dunes.

And so on.

Why do zebras have stripes and aren't gray all over? Because gray uniformity is an unstable state of zebra chemistry, and its symmetry breaks to give stripes.

Why isn't a splash circular? Because that circular state is unstable, so the symmetry breaks to the discrete rotations and reflections of a crown.

By listing the possible subsets of symmetries, we list the possible patterns that arise through symmetry-breaking. In this instance, and in others, we end up listing virtually all the patterns anyone ever sees. In short, symmetry-breaking is a universal mechanism of pattern formation.

Where did the rest of the symmetry go? Again, it gets spread around over lots of different solutions. The dunes could form in any position. In fact, they do, because they slowly creep across the desert. The different positions are related by translational symmetries—precisely the symmetries that are broken to get the pattern in the first place.

The same goes for innumerable other pattern-forming systems—flowers, spiral waves, the ripples of the millipede's legs… And snowflakes.

THE ORIGIN OF SPECIES

Symmetry-breaking can also shed light on an old evolutionary puzzle. Even the most casual observer is likely to notice that although animals and plants are enormously diverse, they occur in clusters—taxonomic, not geographical. These clusters of very similar organisms are called species.

Over the timescale of a human life, species seem fixed. In July 1837, however, the naturalist Charles Darwin, having returned from exploring various parts of the world on board the *Beagle*, opened his first notebook on *Transmutation of Species*. Darwin had begun to suspect that over very long periods of time species can change. He formed this opinion by observing related species in isolated regions. An important example, for him, was a set of 13 species of finch on the Galápagos Islands. It looked as if they all diverged from a common ancestor, probably a small group of finches of a single species blown to the islands in a storm.

Darwin's epic work *On the Origin of Species* explained such changes as the gradual accumulation of small, inheritable differences, brought about by natural selection. Nowadays we use the word "evolution." Natural selection is often described by a slogan, "survival of the fittest," but the crucial point is not fitness— it is that some creatures survive to breed and others do not. This difference can offer a selective advantage for certain changes in the way the organism behaves or how its body is built. If so, these changes will prevail in succeeding generations.

We now have extensive and definitive evidence, including the fossil record and genetic sequences in DNA, that species have evolved. Once, about five million years ago, the distant ancestors of chimpanzees and humans were all part of a single species. Today they are not.

How did the species diverge? Was Darwin right to think it was a gradual drifting apart? But if there was a selective advantage in changing, why didn't they all drift in that advantageous direction? Why did some go the other way?

This is not the only problem. Species, by definition, consist of organisms that potentially can interbreed. The biologist Ernst Mayr pointed out long ago that interbreeding mixes up the genes and acts as an obstacle to species divergence. His proposal: something prevents the nascent species from interbreeding. According to his allopatric (different fatherland) theory, a small group occasionally becomes separated from the main population by a major geographical hurdle—a mountain range, a big lake—and evolves for millions of years without contact with the main group. By the time the two groups get back together, their independent evolution has changed them so that they no longer interbreed, even potentially.

This theory is now being challenged by a less intuitive idea—sympatric (same fatherland) speciation. Biologists have proposed many mechanisms to keep the nascent species apart— such as sexual selection, in which females come to prefer certain traits in their males and mate selectively with those that have them. Mathematically, all of these mechanisms have a common feature. They view speciation as a

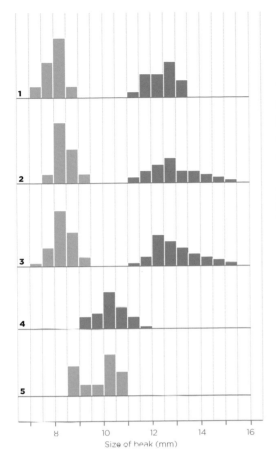

1 Abingdon, Bildloe, James, Jervis

2 Albermarle, Indefatigable

3 Charles, Chatham

4 Daphne

5 Crossman

Geospiza fuliginosa

Geospiza fortis

Size of beak (mm)

BELOW LEFT AND LEFT
Charles Darwin studied
the formation of new
species in the finches
of the Galapagos Islands
(below left). The sizes
of finch beaks change
if a species is on its own,
as on Crossman and

Daphne islands, instead
of living alongside other
species, as on Charles and
Chatham islands (see the
graph on the left). These
changes support the theory
that speciation is a form
of symmetry breaking.

symmetry-breaking bifurcation. A single species is a highly symmetric system—all of its organisms are effectively interchangeable. A cat's a cat, and for many purposes it doesn't much matter which cat; ditto mice. A system of two distinct species is less symmetric—change a cat for a mouse, and it really does make a difference.

Mathematical models of speciation as a symmetry-breaking bifurcation lead to some surprising but very general predictions. The first is that this kind of speciation event is very sudden, quite unlike Darwin's gradual accumulation of tiny changes. Another is that the two species "push each other apart," away from the original common body plan. If a species of birds with medium-sized beaks splits into two, then one group has shorter beaks, the other has longer ones, and suddenly there are very few birds occupying the old middle ground.

How can this happen if the genes are constantly being mixed up? The answer is… natural selection.

This kind of bifurcation occurs when selection stops favoring the middle ground. Maybe there aren't enough seeds any more that are the right size for medium beaks. Maybe the very success of the medium-beaked birds has overstretched the available resources. If so, then selection removes the hybrids formed by interbreeding between a big-beaked bird and a small-beaked one. The genes are there, they still get mixed up—but those combinations are removed before the bird can breed. This kind of selection can be observed today. An excellent example, in fact, is Darwin's finches, which respond astonishingly rapidly to changes in climate and vegetation.

13

FRACTAL GEOMETRY

Sometimes, then, we can learn about nature from mathematics. This is the subject's way of repaying a huge historical debt. Mostly, mathematics has learned from nature. In the 1970s Benoît Mandelbrot, then a scientist working for IBM, became aware that there was a common thread running through his work.

He had been studying all sorts of apparently disconnected problems—the stockmarket, the amount of water in rivers, interference in electronic circuits. The common thread, he suddenly realized, was that each problem had intricate structure at all scales of magnification. If you graph price movements in the stockmarket on a monthly basis, you get a rather irregular curve with lots of ups and downs. If instead you look on a weekly basis, a daily one, an hourly one, or even minute by minute, you still get a rather irregular curve with lots of ups and downs. The same goes for the water flowing in a river, or the changes in current in a noisy electronic circuit. Mandelbrot decided that this kind of structure needed a name, and he invented one—fractal. A fractal is a geometric shape that has fine structure no matter how much you magnify it.

Most of the familiar forms in the natural world are fractals. A tree, for example, has structure on many scales—trunk, bough, limb, branch, twig. So does a bush, a fern, or a cauliflower. A lump of rock looks like an entire mountain in miniature; a small cloud looks just as complicated as a big one if you view it close up; the surface of the Moon is covered in craters of all sizes.

The traditional shapes of Euclid do not behave like this. There is nothing intricate about a triangle, circle, or sphere. If you magnify a sphere, then its surface becomes flatter and flatter, resembling a featureless plane. But if you magnify a distant mountain—say by walking toward it and getting right up close—you start to notice fine details that were originally hidden from view.

Fractal geometry is a mathematician's idealization. Nature's fractals are extremely intricate, but they fuzz out on the atomic scale. A mathematician's fractals are infinitely intricate and never fuzz out, no matter how closely you look. The ideal is simpler than the reality because you don't have to worry about what happens on the scales where things fuzz out. Coastlines are good examples of fractals. How long is the coast of Australia? On a large-scale map, Australia looks complicated, but you can imagine measuring the total length of the coast as drawn on the map, perhaps by using one of those gadgets that looks like a pen with a little wheel in place of the nib. However, if you change to a larger scale map, you discover lots of bays and promontories that the first map couldn't show because they were too tiny on that scale. If you include these extra wiggles, the measured length grows substantially. The more detailed the map, the longer that coastline seems to get, until the numbers become too large even to contemplate. Australia is doing an excellent job of approximating a mathematician's fractal coastline, whose length is infinite.

A coast is a fractal line. A mountain is a fractal surface, with jagged peaks that are made from smaller jagged peaks that are made from even smaller jagged peaks that are…Clouds are fractal surfaces too, with wisps of water vapor that break up into ever finer wisps the more closely you look. Trees are fractal plants. Rivers are trees of flowing water—the main river is like a trunk, its larger tributaries are branches, the tiny streams up in the hills are the twigs. The water that ends up in the river erodes the land into fantastic treelike patterns. Our entire planet is a fractal, if you are a geologist.

When nature keeps reusing the same catalog of patterns, the wise scientist pays attention. Mathematicians, in particular, have made enormous progress by learning to see the hidden patterns of nature. Even Euclid's triangles were originally derived from the measurement of the land—that's what "geometry" means. So it's worth paying attention to fractals.

BELOW Coastlines look irregular, and the closer you look, the more irregular they become. As the scale of magnification increases, new details start to appear which reveal irregularities that had previously been too small to see. A coastline is a naturally occurring fractal.

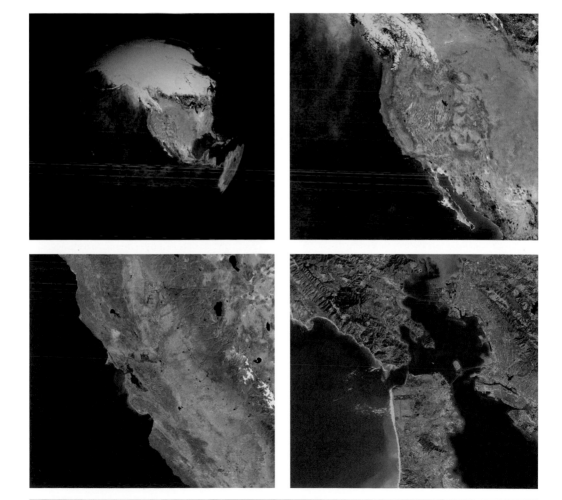

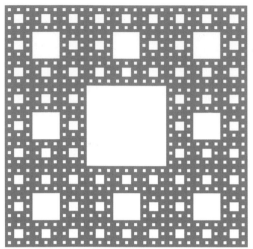

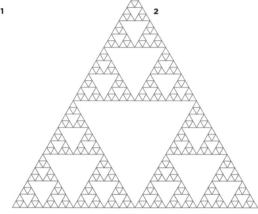

FRACTAL MATHEMATICS

The mathematical key to nature's fractals is the concept of self-similarity. Recall that a shape is self-similar if it is made up from smaller copies of itself (*see pp. 34–35*). Squares are self-similar—64 small squares make one big square chessboard. But squares are old-fashioned, straight out of Euclid—self-similar but not intricate. Can a more complicated shape be self-similar? If it can, intricacy on all scales is guaranteed.

Around the time of World War I Waclaw Sierpinski, a Polish mathematician from Lvov who was interned in Russia during the war, invented several fractals—though in those days they were known as pathological curves, mathematical monsters. They seemed monstrous then, because nobody had spotted the naturally occuring fractal patterns in nature.

One of Sierpinski's fractals is made by starting with a square. Divide it into nine equal squares and remove the central one, leaving a border of eight squares, each one-third the size of the original. Now do the same to each of those eight

smaller squares, and so on, infinitely often. This is the Sierpinski carpet, made from eight copies of itself, each one-third the size.

If you use a triangle instead of a square, you get a Sierpinski gasket. This time the fractal is made from three copies of itself, each half the size of the preceeding.

In the late 19th century, the German mathematician Helge von Koch invented a different fractal based on a triangle. Instead of cutting out a hole, you glue new triangles to the edges. Start with an equilateral triangle. Glue on three smaller triangles, each one third as big, to the middles of the sides, giving a six-pointed star. Now repeat with 12 extra triangles one-ninth as big as the original, then 48 extra triangles $\frac{1}{27}$ as big…You get a mathematician's "coastline" for a sixfold island. The shape has finite area (it fits on the page) but infinite length (each stage in the construction multiplies the length by $\frac{4}{3}$). It also has sixfold symmetry, and it is called (aha!) the snowflake curve. Too regular to resemble a real snowflake, it nevertheless captures something of the snowflake's treelike form.

3

LEFT Here are some mathematicians' favorite fractals. Each is formed by repeating a geometric construction over and over again to infinity, on ever smaller scales. The Sierpinski carpet (1) is created by taking a square and making a square hole one third as big in the middle. Then eight smaller holes are made in the remaining subsquares, and so on. A similar process using triangles leads to the Sierpinski gasket (2). Six successive stages of the snowflake curve are shown in one picture (3), reading clockwise from top right. At each stage, a tiny triangle is added to every edge. The resulting shape has a finite area but an infinite perimeter.

Closely related, but "inside out," is the flowsnake fractal. Mathematicians love spoonerisms and other wordplay. Finally, I'll mention the Menger sponge, created in the same way as the Sierpinski gasket, but using a cube. A Menger sponge is made from 20 copies of itself, each one-third the size.

The variety of fractals is, literally, infinite. Some fractals look denser, "rougher," than others—they seem able to fill out more space. Associated with every fractal is a number, its fractal dimension. And the rougher the fractal, the bigger its corresponding fractal dimension.

Ordinarily, we think of a dimension as a direction in space. A line is one-dimensional, a square two-dimensional, a cube three-dimensional. The dimension of a fractal need not be a whole number. For instance the fractal dimension of von Koch's snowflake curve is close to one and a quarter. Does the snowflake poke out in one and a quarter different directions?

Not a bit of it. It's just one of the mathematical fraternity's bad habits—reusing an old name in a new, more general context. Fractal dimension yields the same numbers as the more usual concept of dimension for familiar spaces like a line, square, or cube—namely 1, 2, and 3—but it applies to more shapes. What fractal dimension does is to capture how the number of building blocks relates to their sizes.

Look at the snowflake curve—or, more precisely, one-third of it, the part that runs along one edge of the original triangle. Four copies of that curve make a curve exactly the same shape but three times as big. If the curve was like a line (one-dimensional) we'd need three copies—too small. If it was like a square (two-dimensional) we'd need nine copies to do that—too big. So the dimension of the snowflake ought to be somewhere between 1 and 2 (and closer to 1 than 2, in all likelihood). You can't do that with whole numbers, but one and a quarter fits the bill nicely. There's a technical definition for the fractal of the snowflake curve, and it leads to a value of 1.2618.

SIMPLE RULES

"Theories," the biologist Peter Medawar once said, "destroy facts." He wasn't objecting to facts, and he wasn't really saying that the facts somehow disappeared when a theory was found to explain them. What he meant was that a good theory removes the need to know huge lists of apparently unrelated facts, by reducing them to something simpler.

Think of the solar system. Before Newton discovered his law of gravity, an accurate description of what was known about the solar system required the construction of vast tables of planetary positions over long periods of time. After Newton, all you really needed to know was the positions (and velocities) of the planets at some chosen instant, plus the laws that governed their subsequent (and previous) motion. Then you could work out all other positions from the laws.

A more modern version of Medawar's statement is "Rules compress data." If you specify an initial state and a dynamic process—which on a computer consists of a list of numbers— you can generate as many numbers as you like. When you buy a computer nowadays you generally get a lot of free bundled software. Among the demos and games you often find a complete encyclopedia on a single CD-ROM. You may, in passing, marvel at the technological advances required to cram so much information into so small a space. But there's more to it than just technology. There's new mathematics, too, which makes it possible to cut the amount of information down to a manageable size before the technology even begins to cram it in. Storing text is easy, it's images that cause the trouble. One popular CD-ROM encyclopedia, for instance, includes 8,000 color pictures. The information in a picture can take up as much space as a hundred pages of text. A screen image is made

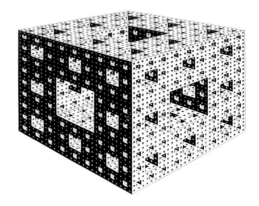

ABOVE Yet another mathematical fractal, the Menger sponge. This is constructed like a Sierpinski carpet, but using cubes instead of squares. All of these fractals are self-similar—made from parts that have the same form as the whole.

up of millions of tiny pixels, individual dots of color. Today's technology cannot store 8,000 images on one CD-ROM if they are represented in this pixel-based fashion. But there they are, shown in exquisite detail, in the encyclopedia.

How do you compress a picture? The most important message from fractals is that simple rules can produce complex shapes. Michael Barnsley, a British-born mathematician living in the United States, realized that this peculiarity of fractals might form the basis of a method of image compression. The key example that drove his thoughts was a fern. A fern consists of a large number of fronds arranged two by two along a stem. Moreover, each frond is like a miniature fern, having its own frondlets, and so on. The entire structure of the fern can therefore be

ABOVE A beautiful self-similar structure in nature is the fern we encountered earlier, shown again above. Every frond closely resembles the entire fern. In nature, however, the similarity ceases at large enough scales of magnification. In mathematical models, it goes on forever.

represented by a collage in which the original fern is dissected into four "transformed"—distorted—copies of itself. These copies are the two largest fronds, near the base of the stem; the rest of the fern above these fronds; and the tiny bit of stem left at the base—which, in a bit of a cheat, is thought of as a fern squashed flat.

Like the old Chinese recipe that begins "Boil twenty chickens in a pot, then throw away the chickens," the trick is to throw away the fern. Barnsley found a method to regenerate the fern, in full detail, from the mathematical transformations (known as an iterated function scheme) which determines the collage.

Each transformation can be specified by just six numbers, so 24 numbers are all you need to generate the incredibly intricate shape of the fern. And the computer disk need store only those 24 numbers, not the million needed to determine the shape one tedious pixel at a time.

Ferns are all very well, but what about completely general pictures? Can they be similarly broken down into distorted copies? The answer is yes, but you don't use copies of the entire picture. What you do is find tiny bits of the picture that resemble other, larger pieces of it. Given enough such bits, you can again represent the image in terms of a collage of mathematical distortions—and the list of numbers that is needed to specify them is far shorter than a pixel by pixel description would be. With Barnsley's method an image can be compressed to take up less than a hundredth of the usual space.

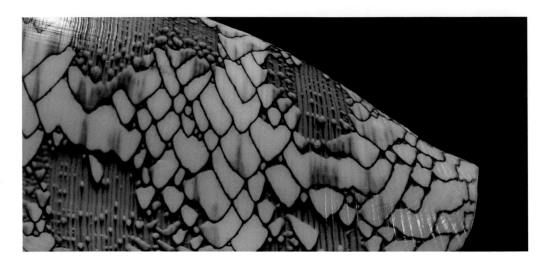

NATURE'S FRACTALS

Fractals are inherently beautiful, but they also yield useful insights into many different questions. Even practical ones. For example the Sierpinski gasket turns out to be a wonderful design for the antenna of a mobile phone. You have to be inordinately insensitive to how scientific progress works to dismiss the fractal geometry of nature as some kind of unimportant coincidence.

The Sierpinski gasket seems to pop up rather a lot. My favorite natural example is in shells. Several different shell species display patterns remarkably similar to the Sierpinski gasket. They include the tent olive shell (found off the eastern coast of South America), the mammal volute shell (South Australia), Damon's volute (Western Australia), the textile cone shell (Indo-Pacific), and the dramatic Glory-of-the-seas (Southwest Pacific). This similarity could be dismissed as coincidental—or unimportant—but it seems to have more significance than that. As we've seen already (*pp. 124–125*), Hans Meinhardt has compiled impressive evidence that the patterns of shells arise from chemical processes in and

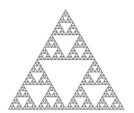

ABOVE AND LEFT Intricate though the fractal markings on a cone shell (above) may be, these are produced by simple rules for deposition of pigment. The shell pattern is similar to a Sierpinski gasket (left)—also a complex pattern produced by simple rules.

around the growing lip of the shell. These processes are in the general category of Turing's reaction-diffusion systems, with an added dash of biological realism; the pattern reflects complex spatio-temporal patterns of reinforcement and inhibition resulting from the interplay of hormones and other genetic products. Such mathematical processes are comprehensible, and we now know that they can give rise to obvious regular patterns, like stripes, or to more subtle ones—like fractals.

So to me—independently of the precise details of Meinhardt's theories, by the way—what the fractals on seashells are telling us is really rather important. It is that the growth patterns of living organisms are consequences of dynamical rules— in this case, applied to networks of chemical reactions. We can sequence genes all we want,

13 FRACTAL GEOMETRY

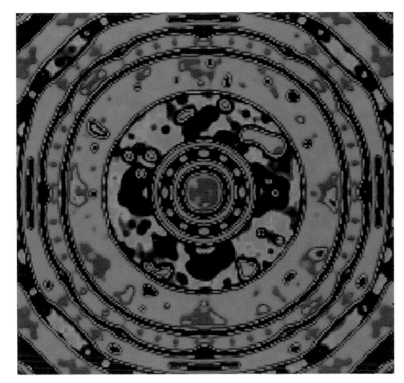

ABOVE AND LEFT The rules for a Sierpinski gasket generalize to cellular automata, which can create complex patterns and are used to model ecosystems.

but it won't tell us what their dynamical interactions do.

We can see the point about dynamical rules clearly in a simple analog of Meinhardt's theory, a cellular automaton. We model the growing edge of the shell as a line of squares, which moves down a grid of squares one step at a time. We model the pigments and hormones by allowing a square to be black or white; we model the dynamics of reaction and diffusion by rules for how the next row of squares depends on the previous one.

For example suppose the rule is "each square is the opposite color to the one above," meaning that each chemical inhibits its own production but enhances the other one. Here we have reaction in each location, but not diffusion—the state of each square depends only on the square that was

in the same position last time. If, for example, the top row is all black, then the next is all white, then all black again, and we create a pattern of alternating black and white stripes. Here a repetitive rule gives a repetitive pattern.

Here's a marginally more complex rule, with some built-in diffusion. If the two squares immediately above a square in the new row, to its right and left, have the same color, then the new cell is white. Otherwise, it is black. For maximum effect, start with a single row consisting of black squares sandwiched between white ones (but a random choice of black and white works just as well).

What do we get? A Sierpinski gasket.

Repetitive rules can also give nonrepetitive patterns. Another treasured intuition bites the dust.

THE MANDELBROT SET

Mandelbrot's name will forever be associated with a fractal that he invented and made popular—the Mandelbrot set. Unlike most of his fractals, this one is an exercise in pure mathematics—mathematics for its own sake, for the joy of intellectual discovery.

I see no harm in that. Mathematics results from the interplay between its own internal dynamic (which gets tarred with the name "pure" as if purity is either a virtue or a vice, rather than a description of a useful methodology) and its relation to the external world (which gets tarred with the name "applied" whether or not its application contains anything of the slightest interest to anyone other than its proud inventor). This interplay is vital, and the absence of either component damages all ideological versions of mathematics.

In everyday life we use so-called real numbers, which are either positive or negative. Positive numbers have square roots, negative ones don't. From 1545 onward, mathematicians began to realize that their lives could be made enormously simpler, and intellectually richer, by inventing a new brand of number in which negative numbers do have square roots. They are known as complex numbers, and the main added ingredient is a number rejoicing in the symbol i, which is the square root of minus one. Provided you stop worrying about what this beast is, you can do arithmetic and algebra with complex numbers and never run into any difficulties. Modern concepts of "number" have made the philosophical status of i clear, mostly by showing that the philosophical status of ordinary numbers like 1 and 2 is far more subtle than we assume in our day-to-day dealings such as standing at the supermarket checkout.

Dynamics is about what happens when you repeatedly apply some rule. Mandelbrot decided to take pretty much the simplest interesting rule he could think of: "square and add a constant." This rule was known to produce some fun dynamics even for real numbers. But he added a crucial, creative twist: "apply the rule to complex numbers."

A real number can be visualized as a point on a line—positive numbers to the right, negative ones to the left. In the same way, a complex number can be visualized as a point on a plane—ordinary numbers run horizontally from left to right and multiples of i run from bottom to top. So a dynamical rule for complex numbers describes a point hopping about in the plane. By convention, the dynamical process starts with the number 0.

What happens as we keep applying the rule depends on the value of the constant in "square and add a constant." The constant can be any complex number. For some constants, everything is very boring. As you keep applying the rule, the point in the plane just wanders off to ever more distant reaches, yielding a series of ever more widely spaced dots. For others, though, the point fails to escape, and then it makes astonishingly beautiful and intricate shapes.

The difference between "escape to infinity" and "stay trapped" is a big one. Which constants yield which behavior? The most plausible guess is that there must be some simple way to decide, some obvious property of the constant. Mandelbrot carried out computer experiments to find out. Divide the plane into tiny cells and choose a complex constant at the center of some cell. If the dynamical rule, applied using that constant, leads to an escape, then color that cell white; if the point stays trapped, color it black. Do this for every cell, one at a time. What shape do you get? Is it simple, like a black circle?

Not at all. It is incredibly complicated. It twists and twirls and spins and branches and extrudes lots of blobs that are in turn just as complicated as the whole thing. It is a fractal, the Mandelbrot

BELOW The structure of the Mandelbrot set creates beautiful patterns like this one.

set, and it has a haunting elegance. (It also looks very like a psychedelic hallucination, but let's not dwell on that.) It is absolutely amazing how much complexity can be generated by such a simple rule—high-school algebra. The structure of the Mandelbrot set can be made more visible, and prettier, if the white regions are colored according to the rate at which the point heads off to infinity—any color code will do, just choose one and paint by numbers.

Within the Mandelbrot set are shapes like spirals, shapes like seahorses, shapes like trees. There are endless variations on common themes, each variation subtly different from the rest. There are even microscopic copies of the entire Mandelbrot set, perfect in every detail. So the intricacy really does go on forever.

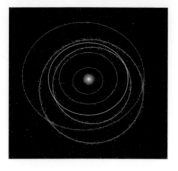

ORDER WITH DISORDER

Fractals are complex shapes generated by simple rules and they betray their origins by combining intricate "disorder" with ordered texture. Snowflakes are complex shapes, too, and we expect them to be generated by simple rules—laws of physics. They also have that characteristic combination of order and disorder, with the order being sixfold symmetry and the disorder the complicated fernlike branching patterns. Is a snowflake a fractal? And if it is, what does this tell us?

A fractal is a mathematical abstraction. A snowflake is a real object. These two things are necessarily different, so a snowflake isn't a fractal. End of story? No—though I'm constantly amazed by intelligent people who think it is. Fractal geometry is controversial—I presume because of its novelty—and the difference between mathematics and reality is a standard argument trotted out to dismiss it. But by the same token, planets are not spheres or point masses, so Newton's law of gravity says nothing about planets. A crystal is not a perfectly regular lattice, so the crystallographic symmetries tell us nothing about crystals. A Nautilus is not a spiral, DNA is not a double helix, and a mirror does not reflect…

Let's put our brains in gear. Mathematical concepts are always idealizations of the real world, not realities themselves. Fine. That's the whole point. That's how we use mathematics to understand nature, it's why mathematics works. We replace the messiness of the real world by a carefully selected idealization, simple enough to be understood—then we have a chance of getting somewhere. So our question becomes: can a fractal idealization give us worthwhile insights into snowflakes? And the answer is a resounding yes.

Over the last 20 years we've learned that many growth processes give rise to fractal forms, which often yield insights into analogous processes, forms, and patterns in the real world. A good example is soot. Soot is a soft, fluffy deposit of carbon particles and other molecules that build up in a chimney, arising from the smoke created by burning coal and wood. Under a microscope, we can see why soot is fluffy. The particles stick together in an intricate, complex, loosely connected structure. It looks like a random fractal. The mathematical model for soot is called diffusion limited aggregation.

A computer generates lots of tiny circular disks, representing carbon particles. They drift around at random until they hit a growing cluster of disks, and when they hit, they stick. The results are convincingly fractal; moreover, they have the same fractal dimension as soot. So there is a quantitative confirmation that the mathematical idealization isn't too unreasonable.

Similar ideas have been applied to the spread of steam injected into an oilfield to push out otherwise inaccessible deposits of oil and to the deposition of thin layers of gold on flat surfaces, important in the electronics industry. The fernlike branching often seen in snowflakes can be modeled by a similar, but more regular,

growth process. An ice crystal grows by accumulating new molecules of water on its surface. (In a storm cloud, these molecules condense to form supercooled water vapor.) The growth of the surface, and its changes in shape, can be analyzed using mathematical models. The main cause of fernlike patterns is a phenomenon known as tip splitting. Certain combinations of humidity and temperature create conditions in which flat surfaces are dynamically unstable. If a flat surface develops a bump, then the bump grows faster than other regions on the surface. So the bump gets bigger and bigger…but a big, rounded bump is nearly flat (like a blister on wallpaper), so when the bump gets big enough it becomes unstable too, and new smaller bumps proliferate. It's like a growing plant shoot whose tip repeatedly splits into two or more smaller shoots—hence the name tip splitting. Mathematically, the whole process can be seen as a cascade of symmetry-breaking events— breaking the translational symmetry of a flat surface. The result, in the case of a growing ice crystal, is a fernlike pattern known as a dendrite.

So there was a reason why that frost on my childhood windowpane looked like a forest. And the snowflake, for some purposes at least, can usefully be deemed a fractal.

LEFT If we want to understand the snowflake more completely, it makes sense to abstract its simple features too. So just for the purposes of mathematical modeling we assume that it has perfect sixfold symmetry and displays fractal branching patterns.

THE FRACTAL UNIVERSE

The urge to get really ambitious is becoming irresistible. Let's go after bigger game. What shape is the universe? How is matter distributed within it?

In Newton's day the universe was thought to be of infinite extent—in effect, it had no particular shape. It was just a standard mathematical three-dimensional space, the same one at all instants of time.

After Einstein, physicists came to believe that the universe was of finite extent. In fact, it was a very big sphere. And they also thought that on reasonably large scales, the matter in the universe was distributed smoothly. To be sure, in small regions you might find a vacuum or a star—either no matter or an awful lot—so it wasn't the same everywhere; but in really big regions, say thousands of light-years across, you always got much the same quantity of matter. So as well as being spherical, the universe was essentially smooth. The smoothness now looks decidedly dubious. (The spherical shape is also coming into question, but I'll reserve that story for later—*see Chapter 15*.)

Early surveys of the sky found that there were much the same number of stars in every direction, except for the Milky Way where the stars were far denser. As the power of our telescopes increased, we began to realize that matter in the universe is clumpy. The Milky Way is a galaxy, a huge cluster of stars that includes our own Sun. Out in the depths of space are many more galaxies, billions upon billions of them. In between the galaxies, there are very few stars—space there is seriously empty.

The clustering tendencies of matter don't stop there. The galaxies themselves clump together in galactic clusters, and the galactic clusters clump together in supergalactic clusters. Around 1990, the American astronomers Margaret Geller and John Huchra suggested that the distribution of matter in the universe might be fractal rather than smooth, with clumpiness on all scales. They even estimated the universe's fractal dimension. From that point on, there has been a continual battle between two points of view. Advocates of a smooth universe generally maintain that on the largest scales the clumpiness smoothes out. Advocates of a fractal universe then develop new observational techniques to map the distribution of matter on those larger scales, and—lo and behold—they find clumps.

Some physicists find this worrying. According to the second law of thermodynamics, matter should, in the long-run, smooth itself out into a kind of lukewarm soup, known as the "heat death of the universe." The clumpiness of the universe sits uneasily with the second law. Why hasn't the matter spread out evenly, as the second law demands?

Roger Penrose has suggested that the universe's initial state must have been incredibly special for this strange behavior to have occurred. However, the real explanation may be symmetry breaking. It has long been known that an even distribution of gravitating matter is unstable. Clumps tend to grow, not spread out. Gravity causes the uniform state to break symmetry and leads to clumping. Because the gravitational effects turn out to be independent of scale, we expect to see clumping on all scales—a fractal universe. For a while this mechanism seemed to be too slow to explain the observed clumping, but with better models of the early universe, cosmologists' computer simulations increasingly resemble the fractal structure that astronomers see in their instruments.

So why doesn't the second law work here? The second law was originally introduced to explain the behavior of gases. It shows that large collections of gas molecules, bouncing off each other, should spread out evenly. The forces

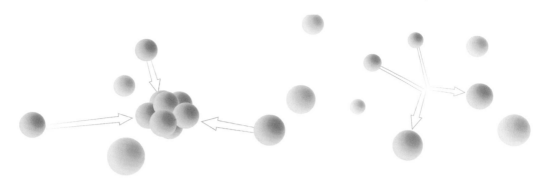

between colliding molecules in a gas are short-range and repulsive. When they hit, they bounce, otherwise they ignore each other. The forces between gravitating particles are the exact opposite—long-range and attractive. Every particle attracts every other particle. The second law is a consequence of the structure of the forces assumed—short-range repulsive forces cause clumps to smooth out because particles are more likely to collide when in a clump. Gravitating systems work the other way. Their long-range attractive forces favor clumps and disrupt evenness. There has never been any reason—other than habit—to expect the second law of thermodynamics to apply to a gravitating system of particles. Our universe is outside its jurisdiction.

TOP AND ABOVE Molecules of a gas interact only when they collide, and then they bounce off each other (top right). The result is a smooth distribution of matter. Under gravity, matter behaves very differently. Particles always interact, and they attract each other (top left). The result is clumpy, not smooth (above).

14

ORDER IN CHAOS

The human mind is an indefatigable pattern seeker. In order to survive in a hostile world, we have evolved a sensitivity to patterns, which we use to predict what's going to happen to us. Even when the world seems patternless, we seek for some rationale to explain it. Sometimes we find that the lack of pattern is an illusion— what looks complicated and messy is actually following simple rules. Sometimes, though, it is our cherished patterns that prove to be the illusions.

BELOW Spurious order in the heavens. Not only is the Bear not a bear—it's not even a real cluster of stars.

For instance our distant ancestors looked at the random scattering of stars in the sky, and their visual systems grouped them into constellations, recognizable clumps—a princess, a lion, a bear. They told stories to their children about the shapes they saw in the heavens. But there isn't really a bear in the sky. There isn't even a constellation, in any meaningful sense. Some of the stars in the Great Bear are only a few light-years away, others are hundreds. If you looked at the stars from a distant world, the shapes and groupings would all seem different. The Great Bear isn't a real pattern, it's not a significant way to think about stars. It has no deep meaning, it teaches us nothing about the way the universe works. But human beings like inventing patterns. We see animals and people among the stars in the sky and weave legends around them to impose a semblance of order on a senseless, random world.

Sometimes, however, the patterns really are there, and they do tell us important truths about the universe. We call these patterns laws of nature. That's what science is all about—digging out the secret patterns that make the universe tick. The most effective way for human beings to think about patterns is to use mathematics.

So we've convinced ourselves that that the laws of nature are mathematical. "God is a mathematician," as the great English physicist Paul Dirac said in 1939.

This view of the role and methodology of science developed slowly, over many centuries, and was around even before there was a recognized concept of "science" as such. It crystallized, in a kind of cultural phase transition, with the work of Isaac Newton. Building on the ideas of many predecessors— "standing on the shoulders of giants," as he put it, with uncharacteristic modesty—he was able to derive mathematical rules for the motion of matter and the action of gravity. All of the complex gyrations of the planets, down to the really tiny details like how the Moon wobbles on its axis and how Jupiter and Saturn alternately surge ahead and tug each other back—all of these things followed from Newton's laws.

The Newtonian mathematics possessed a striking feature, whose philosophical implications took a while to surface. It predicted

the future movement of bodies in terms of their present states. If you know where everything in the solar system is now, and how fast it is moving now, then you just have to crank the mathematical handle to work out where everything will be, and how fast it will be going, a year from now. Crank the same handle again, and you know what will be happening two years from now. Crank it a million times, and you are predicting the future a million years ahead.

The French mathematician Pierre-Simon de Laplace, in the late 18th century, put it eloquently: "An intellect which at any given moment knew all the forces that animate nature and the mutual positions of the beings that comprise it, if this intellect were vast enough to submit its data to analysis, could condense into a single formula the movement of the greatest bodies of the universe and that of the lightest atom: for such an intellect nothing could be uncertain, and the future just like the past would be present before its eyes."

This is the philosophy of determinism— the universe is a clockwork machine and its entire future was determined the instant it was set moving. Of course, Laplace wasn't saying that mere humans could actually do the sums required to predict the future of the entire

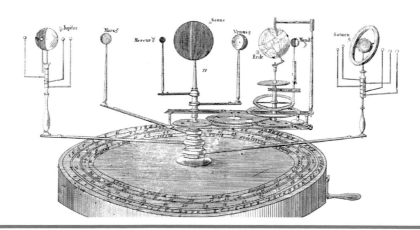

universe—and even if they could, would the clockwork universe let them? Was that action present in their predetermined future at all? How does determinism fit with our apparent free will?

The philosophy, then, was, and indeed still is, deep and difficult to understand. But the science that it gave rise to was impressive and wildly successful. In effect it transformed the entire planet. It changed how human beings think about their place in the universe. It even—in the right circumstances—enabled us to foretell the future.

CHAOS

If you try to take Laplace's eloquent statement as the basis for a program of research, it's easy to home in on the predictive step itself as the most obvious difficulty. As Laplace himself said: "…if it could submit its data to analysis…" It turns out, however, that there is a less obvious difficulty, and this is far more serious because it affects reasonable-scale scientific programs, as well as Laplace's rather megalomanic one. The big problem is not carrying out the calculations that would unfold the future consequences of the system's current state. It is knowing what the current state is.

The first inkling that this is where the real problem lies came about because in 1887 King Oscar II of Sweden offered a prize for a mathematical proof to answer the questions: "Is the solar system stable? Will a planet collide with another or escape from the company of its fellows altogether?" Oscar hoped that the answer was yes and offered 2,500 crowns to anyone who could prove he was right. The French mathematician, astronomer, and philosopher Henri Poincaré took up the challenge and won the prize, even though the problem turned out to be much too difficult even for him. He submitted

a manuscript demonstrating that the motion of a mini-solar system of just three bodies would always follow regular paths.

However, if you go to a library and read the published version of Poincaré's prize-winning memoir, you won't find any such statement about three bodies. Instead, you'll find a statement that is much more interesting—the paths can be highly irregular, far too complicated to predict. Only recently did the history of all this become clear. Poincaré's original submission did claim to prove that the bodies always follow regular paths. But after he had won the prize and his memoir was being printed, Poincaré spotted a mistake. The printed copies were hastily withdrawn, and after a lot of work a corrected version was published at Poincaré's expense. It cost him a lot more than the prize was worth. But scientifically, the revised version is priceless, because in it Poincaré reveals a major new phenomenon. Simple, deterministic equations can have complex, apparently random solutions. We now call such behavior "chaos"—the full phrase should be "deterministic chaos," but the short version packs more punch.

The importance of chaos was not recognized in Poincaré's day. To him, it looked like an insuperable barrier to progress in celestial mechanics, and he stopped working on the problem. His successors, notably George Birkhoff in the 1930s and Steven Smale in the 1960s, took up the challenge and revealed the secret patterns behind Poincaré's chaos. By the 1980s, examples of chaotic dynamics were showing up in every area of science, from astronomy to zoology. Extensive mathematical theories were explaining how chaos arises and why it is possible even when a system is deterministic.

Think of an egg whisk beating egg-white in a bowl. The blades of the whisk go around and around in a regular, predictable pattern, no surprises there. The motion of the egg is much

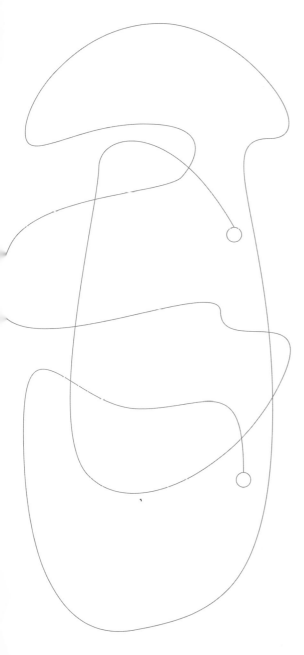

more complex. How do we know that? Because the egg-white gets mixed up. That's what an egg whisk is for. So, what path will a given particle of egg follow? It's impossible to say. Whatever prediction we make, particles indistinguishably close to one another must follow very different paths, ending up somewhere quite different in the bowl. Use food color to paint half the egg red and the other half white. Whisk to a uniform pink. Where does the red half go? Everywhere. Where does the white half go? Everywhere. Now we see the real flaw in Laplace's reasoning. In order to predict where the egg will go, we need to know exactly where it starts—accurate to thousands of decimal places. The slightest error in measuring the starting position will quickly translate into a big error in the predicted motion. But errors of measurement are unavoidable.

As I said, the big problem is knowing where you start from.

Poincaré's discovery boils down to this—the dynamics of a three-body solar system mixes things up much like an egg whisk. Particles that start very close together end up far apart. The motion is deterministic, but this only carries practical consequences if you can measure positions exactly—and you can't. So determinism is not the same as predictability. And Laplace's "vast intellect" has to be able to do more than just submit its data to analysis.

It has to be able to obtain the data in the first place.

LEFT How ordered movements can create apparent disorder; if you beat an egg with a whisk, then the whisk moves in a regular manner, but the egg-white is stirred up in such a complicated way that each tiny bit of egg seems to move at random. The motion is actually governed by laws, but it certainly doesn't look that way.

RANDOMNESS & RULES

The problem with a word like "chaos" is that it is easily misunderstood. Especially when that qualifying "deterministic" is dropped. So it's often thought that "chaos" is just a fancy new word for randomness.

Not so.

Chaos is apparent randomness with a purely deterministic cause. Unruly behavior governed entirely by rules. Chaos inhabits the twilight zone between regularity and randomness. Precisely because chaos runs counter to many of our cherished intuitions, it's not so easy to get the right idea here. For instance it looks as if the easy way out is that word "apparent." Yes, of course, chaos looks like randomness, but it's not,

not really. How can it be when it's produced by rules?

Unfortunately, it's more subtle than that. In certain respects, there is true randomness in chaos. Roughly speaking, the rules of a chaotic system latch on to microscopic randomness in the initial conditions and magnify that randomness so that it shows up in large-scale behavior.

The discussion is made more difficult by a philosophical problem: does true randomness really exist? A standard metaphor for randomness, for instance, is the roll of the dice. But dice are cubes, their rolling is governed by deterministic rules. Where does the randomness of dice come from, then? Because we don't know the initial conditions for the dice when we roll

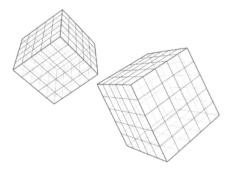

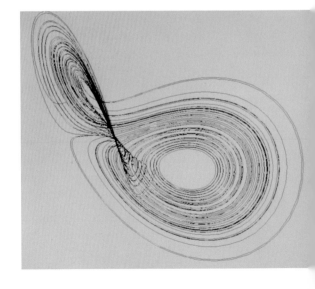

ABOVE AND RIGHT We use dice as a source of random numbers, but in principle dice are just bouncing cubes (above), and obey the laws of mechanics, like other physical objects. Where, then, does the randomness come from? Part of the answer to this is that dice are chaotic.

Chaos may look random, but there are hidden patterns. Associated with chaotic systems are attractors—intricate geometric shapes that capture key features of the dynamics. This Lorenz attractor (right) comes from a chaotic model of weather patterns.

513652315 6632

them—that's why we shake them in our closed hands or in a box—and anyway, even if we did know, the smallest errors in measuring those conditions would be amplified as the dice bounced across the table. Maybe there's only one predetermined way for the dice to land, but neither we, the dice, nor the universe "know" what that predetermined number is until we roll the dice and find out.

There is one possible exception here. Physicists claim that quantum mechanics is based on genuine randomness, so at its smallest scales the universe functions purely according to chance. That may well be the case, though some mavericks think that the probabilistic nature of the quantum world is itself illusory and that it, too, runs on secret deterministic rules. I think that the concepts "random" and "ordered" only make sense with reference to a man-made mathematical model—I'm not sure we can safely attribute either to the real universe as if it were an absolute distinction.

The puzzling dual ordered/disordered nature of chaos is brought into sharp focus if we follow current mathematical practice and represent dynamics in terms of geometry. Associated with any dynamical system is its own geometric space, known as its phase space, whose coordinates are the variables of the system. An initial condition is a specific set of coordinate values, that is, some point in the phase space. As time passes, the coordinates change, thus obeying the dynamical rule—the initial point moves through the phase space along some curve, or flowline. Each initial point generates its own unique flowline, and the system of all these flowing curves corresponds to the flow of the dynamical system.

In nonchaotic systems, the flowlines home in on something simple—a single point for a steady state, a closed curve for a periodic solution. In chaotic systems, they home in on more complex shapes, called attractors. This doesn't mean that they exert some kind of gravitational force. It means that wherever you start from, pretty soon you're very close to an attractor. So the attractor—or attractors if there are several—defines the long-term behavior of the system.

Chaotic attractors repeatedly stretch the flowlines apart and reinject them back into the same limited region of phase space. The flow can't escape, but because of the stretching it also can't do anything simple and straightforward. So the way the flowlines wiggle around looks unstructured and random, even chaotic. Their movement is unpredictable, for the reasons already discussed—tiny errors in knowing the starting points translate into big errors in future predictions.

Chaotic attractors have such an intricate geometry that they are fractal. They have intricate structure on the finest scales. The easiest way to see why is to run time backward. The attractor itself has structure on a large scale. In backward time, the stretching that occurs in chaos becomes shrinkage instead. So that large-scale structure shrinks to create small-scale structure. The further back we run time, the smaller that structure becomes. So the geometry of attractors unites the concepts of chaos and fractals.

6216224611426

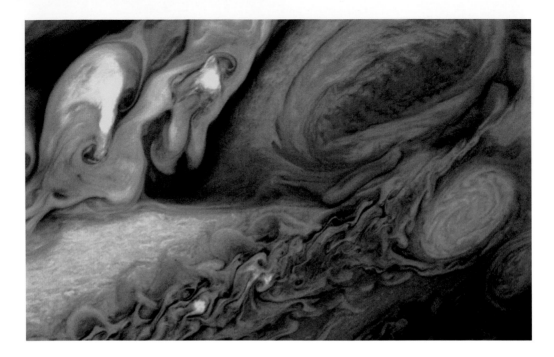

TURBULENCE

Sometimes you can see that chaotic dynamics has some kind of pattern, albeit complex and ever-changing. A good example is Jupiter's Great Red Spot. As we know, this dominant feature of the Jovian cloudscapes is a gigantic vortex, about two Earth diameters long and one wide, which has been twirling away in Jupiter's atmosphere for centuries—probably for hundreds of thousands of years, for it is exceedingly stable. The flow in the spot itself is simple and regular, a spiraling rotation.

When the Voyager spacecraft flew close to Jupiter they discovered that the Red Spot is constantly spinning off a stream of vortices, a wake with definite structure, but a high degree of unpredictability. No one could claim the wake is random—there are clear wave shapes and very characteristic patterns. But those patterns keep changing, on a timescale of a few hours, and no one can predict the exact way they will change.

The turbulent wake arises from interactions between the Red Spot and the surrounding atmosphere. The Red Spot is "stirring" the atmosphere around it. However, the mathematical equations that model the flow of atmospheric gases in that nice, regularly moving Red Spot are the same equations that model the flow of atmospheric gases in that turbulent, unpredictable wake. The equations are deterministic—no built-in randomness. Yet they have two very different kinds of solution. In the Red Spot, the flow is regular. In the wake, it is chaotic. It's just like the egg whisk and the egg.

Turbulence is one of the great unsolved problems of physics. Fluid dynamics is a fascinating area, long studied by mathematicians and physicists, and we know enough about it to make the space shuttle fly (even though it has the aerodynamics of a brick) and to carry millions of people around the world every day on jet airliners. As usual, however, there is even more that we don't know, and turbulence is the great unknown.

Experimentally, fluids seem to flow in two different ways. Either the flow is smooth or it is a surging mass of twisting, ever-changing vortices. If you turn on a tap, gently, the stream of water

LEFT AND RIGHT Order and chaos combine in a single image. Inside Jupiter's Great Red Spot (left), its atmosphere moves in a regular and ordered way. Outside, the atmosphere spins off in complex trails of vortices (right). The laws of fluid dynamics are the same in both cases—but the implications of those laws are different in different circumstances.

that emerges flows smoothly. If you turn the tap on full, though, you often get a frothy, irregular stream of water. These two types of flow are called laminar and turbulent. Laminar flow is easy to explain—the laws of fluid dynamics are deterministic equations, with nice, smooth solutions. What of turbulence?

In the early days of chaos the mathematicians David Ruelle and Floris Takens suggested that turbulent flow is a manifestation of chaos in the equations of fluid dynamics. This suggestion did not go down well with the fluid dynamics community, which at the time preferred an idea that we now realize was naive. The favored idea in those days, the Landau theory, held that turbulence is an accumulation of more and more periodic motions with lots of different periods; the Ruelle-Takens theory denied that such an accumulation is possible, let alone present in turbulent flow. In retrospect, it turns out that the Ruelle-Takens theory got many details wrong. Nonetheless, its main idea—that turbulence is a manifestation of chaos in the deterministic equations of fluid flow—seems to be spot on. We now know that weak turbulence, of a kind

that can be generated easily in the laboratory, can indeed be traced to chaotic dynamics. Tom Mullin, an applied mathematician, has shown that turbulent Taylor vortices in the Couette-Taylor system—fluid between rotating cylinders (*see pp. 90–91*)—correspond to a chaotic attractor. Strong, "fully developed" turbulence is another matter—even if it is chaos, knowing this doesn't help much in understanding what's going on.

Fractals are the geometry of chaos, so we expect to find fractal structures in turbulent flows. And we do. In fact, such structures were known to the fluid dynamicists long ago, before Ruelle and Takens suggested any connection with chaos. One of the most influential descriptions of turbulence was given by the Russian mathematician Andrei Kolmogorov: a turbulent flow is a cascade of rotational energy from large vortices to ever-smaller ones. This sounds like a fractal, and that's exactly what it is. In fact, it's very much the process that leads from Jupiter's Great Red Spot to that elegant, enigmatic, turbulent wake.

POPULATION DYNAMICS

The theory of chaos is changing the way we think about biology, too. People used to talk about the "balance of nature"—the way, say, foxes keep down the population of rabbits, and the way the availability or otherwise of rabbits limits the population of foxes. The image of a balance reflects an unstated but deeply held belief that nature, if left to itself, will settle down to a steady predictable state. Give or take the odd birth or death, there would always be the same number of foxes preying on the same number of rabbits. It's a comforting concept—sustainability that obviously goes on forever.

Comforting it may be, but nature cares little for human comfort. We are finally starting to recognize that the global ecosystem, if left to itself, need not settle down to anything. Certainly not anything as simple as a steady state. Instead of balance there is constant change. First one species gains ascendance, then this triggers the growth of a different species, then those trigger the collapse of a third…the ecosystem surges hither and yon, marching to the beat of an unseen—and only dimly understood—drummer.

In the 1920s the Italian mathematician Vito Volterra developed equations to model the populations of food-fish and predators (such as sharks) in the Adriatic sea. His model predicted not a balance but periodic cycles. First there would be a population boom in the food-fish. The predators would respond, but with a delay—after all, it takes time to breed baby sharks. Then the shark population would shoot up and they would "overfish" the food supply. The food-fish population would then plunge, and a lot of sharks would die for lack of food. With few predators around, the food-fish population would boom…and the cycle would repeat. There is plenty of evidence that real populations fluctuate, but their cycles seem to be less regular than Volterra's. Until recently it was assumed that the irregularities were always the result of outside disturbances. Then, in 1987, Australian-born mathematical ecologist Robert May pointed out that chaotic dynamics can occur in many standard population models. If the real world were following a chaotic dynamic, then the result could well be irregular cycles—but with all of the irregularity being internally generated by the population itself.

For some reason, the proposal that irregularities in animal populations might be

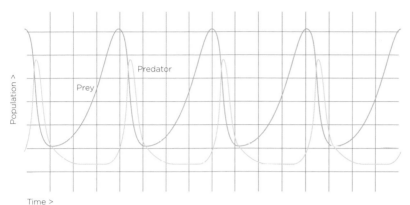

Population >

Predator

Prey

Time >

LEFT Natural populations can undergo complex dynamic changes, even if no outside influences disturb their natural rhythms. One of the simplest models of interactions between predators and prey predicts the occurrence of synchronized cycles, in which the prey population goes up and down, and the predator population does the same after a small time lag.

generated internally by chaos was not welcomed very warmly by the ecological community. Admittedly, it's difficult to test such a theory, precisely because it's difficult to exclude outside influences. Nonetheless, there tended to be an attitude that mathematical models were a good representation of real population dynamics as long as they predicted steady states or periodic cycles, but as soon as those same models predicted chaos, the models became useless. To most mathematicians, who saw the order and the chaos as part and parcel of the same dynamical package, this was baffling. Either you think the equations work, or you don't. When it comes to mathematics you can't pick and choose.

The main experimental difficulty here is to eliminate outside influences. In 1995 Jim Cushing, an American population biologist realized how to do this. Along with some colleagues, he used the technique on a tiny insect, the flour beetle. The flour trade considers these animals to be pests because they get into flour and make it unfit for human consumption. They also have habits that we find unpleasant, for instance they eat their own eggs. Cushing's group formulated a mathematical model of the population dynamics of flour beetles, taking

ABOVE Population dynamics can also be chaotic, so that the size of the population is unpredictable. Are swarms, such as those of bats (see above), natural examples of chaos, are they triggered by external events, or—more likely— a bit of both? We don't yet know, but at least we can ask the question.

these unpleasant habits into account. The model has a number of parameters—death rate, number of eggs laid, and so on. The big problem with external influences is that they might change these parameters. What to do, then? Easy. Make sure that the parameters don't change. If too many beetles die, replenish them with new live ones. If they lay too many eggs, take the excess away.

It's not cheating—it's exactly what a physicist would do to make sure that the temperature or pressure stays constant during an experiment. Purely because of the precise control exercised on it, it becomes possible to test for the presence of the predicted dynamics. Cushing's group carried out the tests, and by 1997 they had found excellent agreement with their model, and the unmistakable fingerprint of chaos—just where they'd expected to find it.

WEATHER FORECASTING

When it comes to prediction, the obvious target is weather. Farmers need to know when to cut the hay or harvest the wheat, and it costs them a fortune if they make the wrong decision. Mountain climbers need to know if a snowstorm is due.

Both weather and tides are governed by laws of nature. These laws are mathematical—and rather similar. The laws for tides describe the motion of a fluid, the oceans, under the influence of the gravitational pulls of the Sun and Moon. The laws for weather describe the motion of a different fluid, the atmosphere, under the influence of the daily cycle of heat from the Sun. We can predict tides years ahead—why not weather too?

In 1922 a scientific maverick by the name of Lewis Fry Richardson published an account of an amazing vision, the weather factory. In his day there were no computers, so he imagined a vast army of people in a building the size of a football stadium, wielding mechanical calculators. The equations for the weather would be turned into a list of calculating instructions. The people would follow these instructions and send each other messages to say what numbers they had obtained in their particular piece of that vast computation. Out of the resulting frenzy would come an accurate prediction of tomorrow's weather, next week's, or next year's.

Of course this mad idea never came to anything, but nowadays we carry out the same task with powerful computers. The frenzy still goes on, as electrons whiz this way and that inside chips and wires. And out of it comes an accurate prediction of tomorrow's weather. The prediction for next week's weather, though, is still often wrong. And predicting next year's weather remains hopeless.

Better computers won't help. And there is nothing wrong with the equations. The problem is, the equations for weather have chaotic solutions. The way the equations work is to start with the best measurements of today's weather, which are obtained from a worldwide network of weather-balloons, ground stations, and satellites. Then the computer runs the rules for weather, to see how these initial conditions will propagate into the future. The trouble is, chaos causes tiny errors in the initial conditions to grow faster and faster, until the errors swamp the prediction.

A meteorologist named Edward Lorenz predicted just this problem in 1963. He came across it when studying an extremely simplified model of atmospheric convection. When he solved his equations on a computer, he found three important things. The first was that the solutions were irregular, almost random. The second was that they could be viewed geometrically as a rather strange shape, which we now recognize as a chaotic attractor (*see pp. 176–177*). (In 2000 the mathematician Warwick

Tucker proved this beyond any shadow of doubt.) And the third was what subsequently became rather famously known as the butterfly effect— the sensitivity of chaotic dynamics to tiny errors. Lorenz gave a lecture in which he said, in effect, that if a butterfly flapped its wings, it could change the weather completely.

We can't test the butterfly effect with a real butterfly because this would involve running the

BELOW Weather patterns can change unpredictably from one month to the next, from one week to the next, or from one day to the next. Advances in computing are unlikely to improve the accuracy of forecasting, because more accurate computations will not be able to overcome the inherent chaos. The prospects for climate prediction may be better: the climate does not depend on fine details.

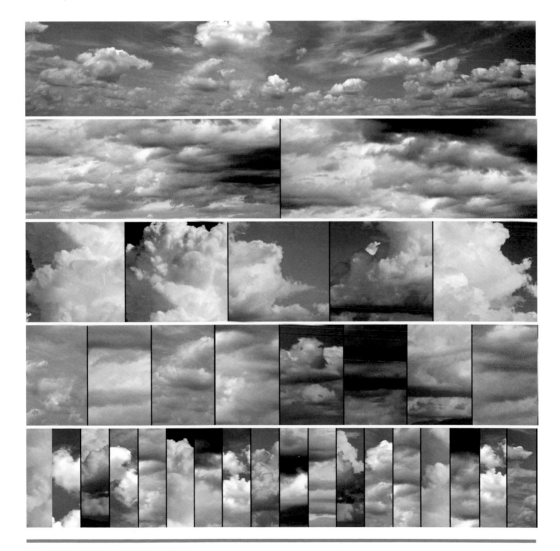

entire Earth's history twice—once without the flap of the wing, and once with it—and comparing the results. But we can test it on the equations that weather forecasters use to predict the weather, and when we do, it turns out that Lorenz was absolutely right. There is a prediction horizon resulting from unavoidable observational errors, and no matter how hard you try, you can't predict the weather any further into the future than that horizon. The prediction horizon seems to be about four to six days, certainly not much more.

The news isn't all bad, though. We can make the chaos work in our favor. Instead of making one prediction, meteorologists now routinely make about 50—one with the real observations and the rest with the same observations after they've been "butterflied," or subjected to small random changes. If all 50 predictions agree about something, then the meteorologists can be confident it will happen. If most of them agree, they can be confident that it is very likely to happen. If the predictions disagree violently, then all bets are off. So now weather forecasts come with an estimate of how reliable they are. In the absence of effective crystal balls, that's the closest to foretelling the future of weather that we can hope for.

CHAOS IN THE SOLAR SYSTEM

Having understood why weather is unpredictable, it pays to take a more critical look at systems that we thought were predictable. The solar system is a case in point. The great triumph of Newtonian physics was the discovery that the solar system obeys simple laws and that it is possible to use these laws to explain the movements of the planets—and to predict them. In 1682 Edmund Halley saw the comet now named after him and successfully predicted its return—he realized that a series of historical records of comets must refer to the same body, repeatedly returning on its elliptical orbit around the Sun, and after that, all he needed to work out was the period of its orbit. Over a hundred years later, the mathematician Carl Friedrich Gauss predicted the reappearance of the first asteroid to be discovered, Ceres, after it disappeared from human view behind the brilliance of the Sun. In 1870 Urbain Le Verrier predicted a new planet, Neptune, and it was found exactly where he said it ought to be. He tracked it down by examining the disturbing effects it had on Jupiter and Saturn.

In contrast, Henri Poincaré discovered chaos in the motion of an artificial three-body solar system (*see pp. 174–175*). Ours has hundreds—surely it shouldn't be simpler? By the 1980s, the scientists who worked in celestial mechanics were starting to wonder whether the solar system was really as predictable as it seemed. The first sign of cracks in the facade of predictability was found in 1984 by astronomers Jack Wisdom, Stanton Peale, and François Mignard, who argued that Saturn's satellite Hyperion should spin chaotically. Hyperion isn't round, it's potato-shaped, and when they analyzed the dynamics of such a body they found that it should tumble chaotically. The orbit was predictably regular, but the direction in which it pointed wasn't. Hyperion had an attitude problem. Subsequent observations, from space probes and from Earth-based telescopes, showed they were right.

Another case of celestial chaos is Saturn's rings. Ever since the Voyager spacecraft flew past Saturn and sent back pictures, we have known that the rings are not just a flat disk with a few circular gaps, as we had previously thought. They are an exceedingly complicated system of narrow rings, packed together like tracks on a CD, with curious gaps and other irregularities. The rings are made from innumerable tiny

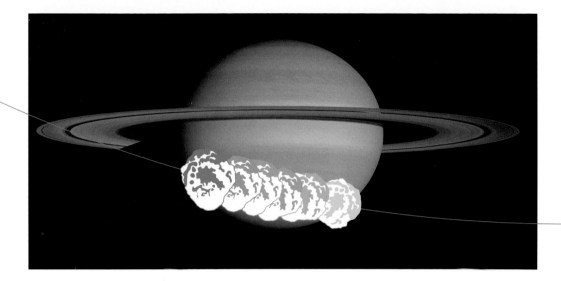

ABOVE Patterns in the heavens led to the discovery of the law of gravity. However, laws can lead to chaos, and our solar system shows many signs that chaos is present. Hyperion follows a regular orbit, but it wobbles chaotically, so it is impossible to predict where each part is pointing. Chaos is also involved in the creation of the gaps in Saturn's rings, and by understanding this relationship, astronomers have been able to predict, and find, new moons.

bodies—rocks, ice. The gaps are swept out because nearby satellites cause these tiny bodies to jiggle around chaotically.

The British astronomer Carl Murray showed that the mathematics of chaos shows a relation between the mass of the satellite and the width of the gap it causes—proportional to the 2/7 of the mass. Some gaps are not known to contain satellites. This formula predicts how massive the missing satellites should be, and gives us a good idea of how likely it is that we can find them with existing telescopes. Several new satellites of Saturn which had the predicted masses were found by this method.

As the years passed, more and more examples of chaos in the solar system came flooding in. Wisdom's group built the Digital Orrery, a special computer dedicated to the sole task of predicting the future of the solar system very, very fast. Running the Digital Orrery to fast-forward the solar system's dynamics, Wisdom's group discovered that Pluto's orbit is chaotic. In a hundred million year's time, it will still be in the same general orbit that it occupies now, but we have no idea which side of the Sun it will be. Since Pluto is part of the solar system and interacts with all the other planets, it came as no surprise to find that on longer timescales still, the entire solar system is chaotic. King Oscar's question about whether the universe is stable is a lot trickier than he thought…

A competing group, headed by Jacques Laskar, has shown that most of the planets share Hyperion's tendency to tumble. They just do it more slowly. Mars, for instance, turns upside down about every ten million years. Interestingly, the Moon stabilizes the Earth, which hardly tumbles at all. Laskar suggests that this may have helped life on Earth to evolve, but this is controversial. His group has also fast-forwarded the solar system for several billion years and it looks like the answer to King Oscar's question is no. Mercury will spiral out slowly until about a billion years from now, when it will have a close encounter with Venus. In all likelihood one or the other will be ejected from the solar system as a result. Which? We don't know.

The solar system is chaotic.

COMET STRIKE!

Mathematics isn't the only thing that links events on Earth to events in the far reaches of the heavens. Our fragile planet is more susceptible to celestial accidents than we like to imagine. The first time anyone looked at the Moon through a telescope, it became clear that our sister world is heavily dotted with craters, some large, some small. For a long time it was believed that these were of volcanic origin, but after the Apollo Moon landings it became undeniable that the majority are the result of meteorite impacts. For billions of years the Moon has been peppered with rocks of all sizes, which have left indelible marks. The same goes for Mercury and Mars, for Jupiter's main satellites, and for all of the asteroids that we've been able to photograph from close up.

Not far from Flagstaff, Arizona, there is a crater 1 mile (1.2km) across and 650ft (200m) deep. It was created in an instant, 50,000 years ago, when a meteorite hit the Earth. Assuming that it was traveling at the average velocity for Earth-crossing asteroids, about 11 miles (17.5km)

per second, the impactor must have been about 500ft (150m) across. For an asteroid, this is tiny. Yet it hit with an explosive force of about 20 megatons. But it was a long time ago and it's highly exceptional. The Earth's thick atmosphere protects us from meteorites, they burn up. But is it all that exceptional? At least 150 impact structures—remains of a meteor crater—have so far been identified on Earth, and there are 30 more whose status is debatable. About the same number presumably struck the oceans, leaving little trace. The main reason we don't see craters all around us is that they have been eroded by wind and rain, while on the Moon and elsewhere little or no erosion takes place. The Manicouagan impact structure in Quebec is visible today only as a ring-shaped lake—but the lake is 45 miles (70km) across. It formed 210 million years ago. Admittedly we don't need to worry about events that long ago, however nasty, but in June 1908 the Tunguska fireball in Siberia flattened trees in a region of tundra about 45 miles square (120km²), and it's pretty certain that this event was a meteorite impact. So they don't all burn up and they're not all millions of years old.

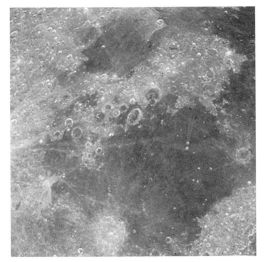

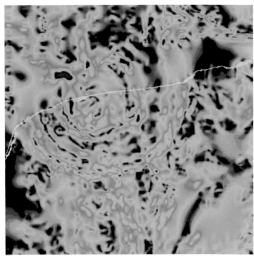

One impact structure has attracted more attention than all the rest put together—at Chicxulub in the Yucatán, on the Mexican coast. It is 110 miles (180km) across. The evidence of its existence is buried beneath subsequent layers of rock, and its presence was first discovered through tiny variations in the Earth's gravitational field. More recently, wells drilled while looking for oil confirmed that it really is there, about 3,000ft (1km) below the surface.

The Chicxulub impact happened 65 million years ago. This coincided with a major change in the Earth's geological and fossil record, the boundary between the Cretaceous and Tertiary periods. Around that time, there was a mass extinction—one of four or more such events that we know of. Innumerable different species of plants and animals died out; most famously among them, of course, were the dinosaurs.

It had always puzzled paleontologists that the dinosaurs, which were an astonishingly successful type of animal that had survived without any apparent problems for around 160 million years, suddenly disappeared from the face of the Earth.

FAR LEFT There are lots of rocks in space, and every so often one of them hits something else. This Arizona meteor crater offers dramatic proof that it could be us. A somewhat larger impact is thought to have helped kill off the dinosaurs.

ABOVE The Moon (left), like all bodies in the solar system with little or no atmosphere is covered in impact craters. The Chicxulub crater (right), one of the largest on Earth, looks even more impressive as an aerial radar image.

In 1980 the American physicist Luis Walter Alvarez discovered in rock layers dating from the Cretaceous-Tertiary boundary a thin layer rich in the unusual element iridium. The Earth has very little iridium normally, but meteorites often have quite a lot. Alvarez concluded that a meteorite about 6 miles (10km) across had hit the Earth. Since then, evidence has piled up to support this idea. Did the K/T meteorite (in German, "Cretaceous" starts with a K) kill off the dinosaurs? If it didn't, it certainly didn't enhance their environment. Its effects would have been considerably worse than a nuclear war, throwing vast amounts of dust into the atmosphere, cutting out the sun, and killing vegetation, with effects that rippled up the food chain. An alternative suggestion is that at the

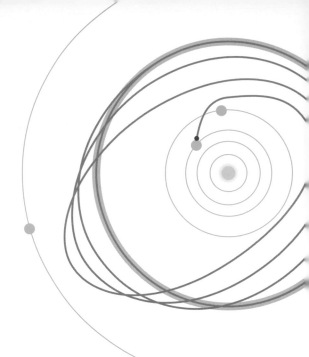

same time there was an enormous amount of volcanic activity on the far side of the Earth from the Yucatán, in the Deccan traps of India.

The most interesting theory of all is that shock waves from the K/T meteorite surged through the Earth and became focused on the opposite side to the impact, triggering volcanoes that created the Deccan traps. In 2015 geologists discovered that the flow of lava from the Deccan traps doubled shortly after the impact. Whatever the details, 65 million years ago the dinosaurs and everything else on the planet were given a dramatic lesson in how events in the heavens can affect life on Earth.

PROTECTOR & DESTROYER

The K/T meteorite could have been an asteroid, or it could have been a comet. If it was a comet, then it came from the Oort cloud, a vast but diffuse collection of a hundred billion would-be comets believed to orbit the Sun, starting at about 25 times the distance of Pluto and extending a third of the way to the nearest star. So the dinosaurs' demise may have been an unpredictable consequence of the workings of chaos among trillions of frozen snowballs. If, instead, the K/T meteorite was an asteroid, then it also fell Earthward because of chaos, but the culprit was a lot closer than the Oort cloud. It was, in all likelihood, Jupiter.

This is ironic, because one of Jupiter's roles in the solar system seems to be keeping comets from getting anywhere near the inner planets. Jupiter is by far the most massive of the planets and it has a correspondingly strong gravitational field, which sweeps up passing comets. The world saw this in a dramatic way in July 1994 when comet Shoemaker-Levy 9 came close to Jupiter, swung past it, broke into 21 fragments and—just as the astronomers had by then predicted—crashed into Jupiter with the energy of ten

thousand hydrogen bombs. Dark marks bigger than the Earth appeared in Jupiter's atmosphere, lasting for weeks. Calculations suggest that this is typical—any comet that comes close to Jupiter is likely to suffer a similar fate.

Asteroids, however, are another matter. Most start from a position well inside Jupiter's orbit and a close encounter with Earth is out of the question. Asteroids are relatively small bodies—the biggest, Ceres, is 290 miles (467km) across, and many are tens of miles in diameter, or less—orbiting the Sun between Mars and Jupiter. Like Saturn's rings, asteroids are not uniformly distributed, but because they are not as dense as the rocks in Saturn's rings, this becomes apparent only when you carry out analysis. The distances of asteroids from the Sun show strange patterns. Some distances are common, so you get clumps, and others don't occur at all, giving gaps.

The clumps and gaps both have the same explanation, and it is essentially the explanation for the gaps in Saturn's rings. At certain distances from the Sun, an asteroid will be in resonance with Jupiter. This means that its orbital period is some simple fraction of Jupiter's period, such as half or two-thirds. An important consequence of resonance is that at regular intervals the asteroid

is in the same position relative to Jupiter, so it always gets tugged the same way by Jupiter's gravity. Depending on the fraction, the effect of this is either to stabilize the asteroid, or—more often—to destabilize it. Chaos is associated with most resonances, and the effect is to push the asteroid out of that orbit (and also out of resonance). A few resonances—the best example is the Hilda group with periods that are two-thirds that of Jupiter—have so little chaos that it doesn't do much damage, and there the resonance causes clumping. Suppose an asteroid starts out in an orbit that is in resonance with Jupiter, in one of the resonances that is subject to chaos. Then the shape of its orbit changes from almost circular to an elongated ellipse. The ellipse can't get long enough for its inner end to get inside the orbit of the Earth—it can't become an "Earth-crossing" asteroid. Lucky? Unfortunately not. It can become a Mars-crossing asteroid. And if it happens to pass close to Mars as it crosses that planet's orbit, then the gravity of Mars can give it a shove, big enough to send it inside the Earth's orbit. If the Earth happens to be in just the wrong place…

Unlikely, yes. But, as we've just seen, some quite big asteroids have hit the Earth, the most

ABOVE FAR LEFT AND CENTER First the bad news…Jupiter can disturb asteroids from their regular orbits around the Sun, and with the help of Mars it can throw them our way (above center). Here's how. Jupiter's gravity disturbs an asteroid. Its orbit then elongates and it breaks free of the asteroid belt. If it passes close to Mars, it may be diverted toward the Earth (above far left).

ABOVE RIGHT Now the good news…Jupiter does, thankfully, an excellent job of sweeping up the incoming comets that might otherwise have hit us. In 1994 the comet Shoemaker-Levy 9 broke up and smashed into Jupiter's dense atmosphere (above right). If it had hit the Earth, the result would have been many times worse than that of a nuclear war.

famous being the K/T meteorite. It may have had some help, it may have triggered volcanic instabilities, it may have just piled the last straw on the triceratops's back—but there's no doubt that an asteroid hit at that time, and it must have made life very difficult for such huge creatures as the dinosaurs.

Yes, Jupiter protects us from comets. But it also flings asteroids part way toward our fragile and precious homeworld, and every so often Mars is in just the right position to head the ball into the goal. And another bunch of animals looks up into the sky and wonders what that bright light is, and…

SYMMETRY & CHAOS

Chaos isn't just doom and disaster. It is intricate structures generated by simple rules—one element that we need to explain snowflakes. But something's missing—symmetry. From the beginning it's been a good bet that an important part of the explanation of the shape of the snowflake must be symmetry. We probably wouldn't have wondered about the shape to begin with if it hadn't been symmetric—an asymmetric snowflake would be little more than an irregular speck of ice.

What we've now discovered, by taking into account innumerable other examples of symmetric patterns in nature, is that the enigmatic regularity of the snowflake is evidence for a much deeper and more extensive symmetry, the symmetry of natural law. In fact, the laws that give rise to snowflakes are considerably more symmetric than the flake itself.

Despite Curie's principle (*see pp. 152–153*), we know that symmetry breaking provides a mechanism for symmetric causes to have less symmetric effects, and it seems likely that the symmetry of a snowflake is what's left of the symmetry of those laws once they've been broken by having to be embodied in a speck of ice. All this explains the regular aspects of a snowflake, but it tells us little about the rest of its form—the irregular aspects. It is precisely the combination of symmetry and irregularity that makes snowflakes so appealing.

So, in a universe of law, where does the irregularity come from? The natural candidate is chaos.

This suggests that we can explain snowflakes if we can somehow combine symmetry and chaos into a single mathematical process. It may sound like combining chalk and cheese, but no matter—they are in fact a lot closer than proverbial usage tries to pretend. Think about

it—both chalk and cheese are animal products, but chalk is the petrified remains of countless dead animals, while cheese has not long squirted from a live one.

In the late 1980s Martin Golubitsky realized that there is an easy way to combine symmetry and chaos in a single mathematical system. A dynamical system is symmetric, as a system, if the laws that it embodies—the rules for determining future from past—are symmetric. That is, symmetrically related causes produce effects related by the same symmetry.

We can translate this rather perplexing recipe into mathematical conditions based on the equations for the system, and by great good fortune the classical mathematicians of the 19th century had already worked out what those conditions imply for the formulas you have to use. So it's easy to find sample equations to experiment with.

Equations without any symmetry of their own sometimes give rise to regular dynamics and sometimes to chaos. It turns out that the same is true for symmetric equations. If you adjust the numbers in the equations in the right way, you can get chaotic dynamics that obey symmetric rules. Symmetric chaos. What do such systems do? The simplest answer comes if you think geometrically and plot their attractors. What you find—and it's fairly obvious, if someone points it out—is that the attractors are chaotic (because the dynamics are) and symmetric (because the rules are).

Write down dynamical equations with the sixfold symmetry of the snowflake, tune the numbers to make the dynamics chaotic, and we can get chaotic attractors with sixfold symmetry. (Some even resemble snowflakes, but presumably that's a visual pun—attractors live in phase space, not real space. Beef up the mathematics a bit, though, and you really can create dendritic patterns through symmetric chaos.) Golubitsky and the French applied mathematician Pascal Chossat, who first played this particular game, called such attractors "icons." The chaos gives rise to beautiful, complicated structures within the attractor, which can be brought out by coloring it according to how likely the system is to visit a particular spot; the symmetry copies the chaos many times over, like a kaleidoscope.

Not long after these discoveries, Golubitsky and I were at a conference on pattern formation where we happened to see a TV program on textiles which often have similar patterns to those on wallpaper, for similar reasons. This made us wonder whether the same ideas could be used to create chaotic wallpaper patterns. The basic idea is identical, but now we use equations with the lattice symmetry of wallpaper—and there are 17 types to choose from. We tried it and hit the bulls-eye on the first attempt, getting a kind of chaotic floral pattern.

What is the significance of a chaotic attractor for physics? It represents a pattern "on average." An example is the Faraday experiment, in which a flat layer of fluid in a dish is vibrated, setting up waves. Even if the container is circular or square, the wave patterns are chaotic. But the averaged patterns have the same symmetries as the circle or square—just as predicted.

RIGHT Symmetry and chaos are not mutually exclusive, but two sides of the same dynamical coin. Mathematically, symmetry and chaos can coexist, and when they do, they create beautiful symmetric attractors. The overall form reflects the symmetry, and the intricate detail is indicative of chaos.

The attractors pictured here are created by simple sixfold-symmetric equations. The resemblance to snowflakes is most probably accidental, but more sophisticated equations do create physically meaningful models of snowflakes from the same ingredients: chaos plus symmetry.

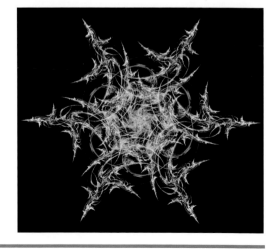

15

LAWS OF NATURE?

One of the great themes that runs through the history of mathematics is how it reveals the unity between human-scale events down on Earth and cosmic-scale events up in the sky. As Newton himself is reliably reported to have said, even though he may have made the story up: gravity acts the same way on an apple and on the Moon. Gravity acts the same way on a galaxy as well, or on a galactic supercluster, or on the whole universe.

The laws of physics, too, are the same throughout the universe. Even though the space between the galaxies is hard vacuum and the interior of Sirius is a nuclear furnace, the laws are the same in both places. What's different is how those laws are being expressed. Transport Sirius out into the intergalactic void, and its interior would still be a nuclear furnace, and it would work pretty much the way it does right now.

Since we're thinking about how the laws that govern a system's behavior affect its form, we may as well go all the way and ask that question about the universe as a whole. What do we known about the form, history, and origin of the universe, and how does this knowledge relate to the laws of physics?

Newton's epic *Principia Mathematica*, published in 1686/87, describes itself as a revelation of "the system of the world." Newton's universe consisted of an absolute space, in which bodies moved, and an absolute time, which determined the changes that occurred. A given moment in time occurred simultaneously at every place in the universe.

At the beginning of the 20th century, Albert Einstein realized that this kind of absolutism is inconsistent with the physics of electricity,

magnetism, and light. He also insisted as an unbreakable principle that the universe must be symmetric—the laws that it obeys must be the same in any frame of reference, even a moving one, as long as it is moving at a constant speed. By combining this "relativistic" principle with a new approach to gravity, he came up with general relativity. According to this, at first sight bizarre, theory, space and time are to some extent interchangeable. And gravity is not a force at all—it is merely a manifestation of bent space-time. The space and time around a massive star are "curved." Einstein wrote equations to describe this curvature. In the early days of general relativity, the only solutions of Einstein's equations that anyone could calculate were ones with spherical symmetry. They led to three types of solution—a sphere that contracted as time passed, a sphere that stayed the same size, or a sphere that expanded. When solid evidence for the expansion of the universe was found by the American astronomer Edwin Hubble, based on observations of shifts in the wavelengths of spectral lines in distant stars, it seemed obvious that the universe was an expanding sphere.

We're no longer convinced it's a sphere, but we're very sure indeed that it is and has been

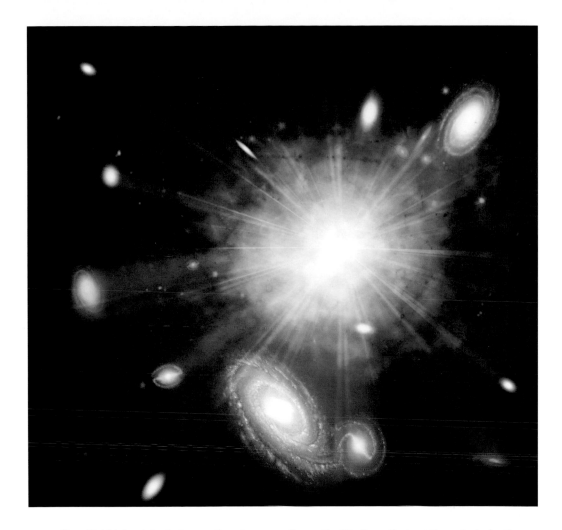

expanding. Hubble's original observations have been confirmed a thousand times over. Running the expansion backward, the entire universe shrinks to a point and disappears. Restoring the normal direction of time, we deduce that about 12 billion years ago the universe came into existence, from nothing, as a single point, which expanded with enormous speed to create today's gigantic universe. This is the Big Bang.

It's important, if difficult, to understand that what expanded was not some kind of bubble filled with time and matter in a surrounding space. The bubble was the space. Space itself expanded from nothing. And time simply was not running before the Big Bang happened.

It was the Big Bang that got time started—the word "before" doesn't apply because there was no before. Not surprisingly the Big Bang theory was exceedingly controversial. Today it has become the consensus view among cosmologists. Because light has a finite speed, the further into space our telescopes can look, the further back in time they can see. So we can check whether some of the earlier stages of our universe are consistent with the Big Bang theory. In particular, we can detect the cosmic background radiation—the smeared out echoes of the Big Bang itself. As far as anyone can tell, these echoes are exactly what we would expect them to be if they really are relics of the Big Bang.

However, some recent observations have required modifications to the original Big Bang theory, and others remain puzzling.

WHAT SHAPE IS THE UNIVERSE?

Symmetry makes equations easier to solve by reducing their complexity. In three dimensions, spherical symmetry is about as symmetric as you can get while maintaining any significant structure. So it comes as no surprise that when physicists first tried to find solutions to Einstein's equations for the shape and dynamics of the entire universe, the only ones they could handle had spherical symmetry. By the time other solutions, with less symmetry, became available, everyone had got so used to thinking about a spherically symmetric universe that it took a while for it to sink in that there might be other possibilities. So although we have gathered a lot of evidence for the Big Bang, we currently have little real evidence for what shape our own universe is.

"Shape" here is not quite the right word to use. We can see what shape an object is by standing at the right distance and looking at it. It sits in its surrounding three-dimensional space and occupies some part of that space. "Shape" is a way of describing how that portion of space taken up by the object relates to the whole. But the universe is the whole. And we're inside it and we can't get outside it to look at the universe from a distance. There is no out and there is no distance to look at it from.

Mathematicians long ago came to terms with this kind of difficulty. Provided we can sensibly talk of the distances between points in some mathematical "object" or "space," then it has a kind of shape—a summary of how all the different points are related to each other. This shape is an intrinsic shape, which is unaffected by how the object is situated in any surrounding space.

A sheet of graph paper laid on a table is flat. One way to tell this is to look at the grid of lines that is printed on it. There are two sets of parallel lines, crossing at right angles. Now pick up the paper and let it bend, forming an arched surface. From the outside, we would say that the paper is curved. From the inside, though, it's still flat. The distances between points, measured along the paper, haven't changed. Take away the

BELOW According to Einstein, gravity is not a force, but an effect of the curvature of space-time. A plane is flat, and as far as mathematicians are concerned it remains flat if it is rolled up into a cylinder. A sphere has positive curvature, and a funnel (modeling the gravitational field of a star) has negative curvature.

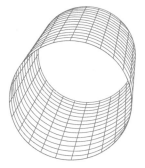

ABOVE We can observe the curvature of space-time by measuring how gravity distorts the light from distant galaxies. The distortion is normally very small, but near a Black Hole it would be enormous. This artist's concept illustrates a supermassive black hole with millions to billions times the mass of our sun.

surrounding space, and all that is left is the internal, intrinsic geometry. If a sheet of paper has that kind of grid, it's flat.

The grid of lines of latitude and longitude on a sphere is a different matter altogether. These lines also cross at right angles, but they are circles, not infinitely long lines. A sphere is not flat, and you can tell this from its intrinsic geometry. Grid lines on a sheet of graph paper and lines of longitude on a sphere have a special property. They represent the shortest paths between their end points. Such paths are called geodesics. So you can tell what intrinsic shape a space has by looking at the geometry of its geodesics. Using this information gives us some chance of finding out what shape our universe really is. So what are the geodesics of our universe? They are the paths along which light rays travel.

When we look at a distant star, we look along one of the universe's geodesics. So if space-time is intrinsically curved, then we ought to be able to see that it is—and we can. The gravity of a galaxy is strong enough to put an appreciable bend into passing rays of light. If there is a very bright light source—another galaxy or a quasar—on the far side of that galaxy, then the distortions of the light can give rise to several images of the same distant object, an effect known as gravitational lensing. Gravitational lensing has been detected in many different parts of the sky, so we have seen the curvature of space-time.

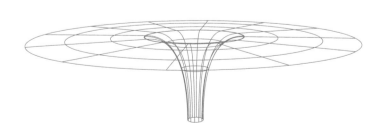

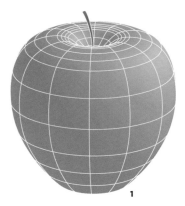

More drastic curvature effects occur when a star gets so massive that it collapses under its own gravitational field. If it is massive enough, it collapses to something so dense that light is unable to escape from it at all. This is a black hole, a region of space-time that has become causally disconnected from the rest of the universe. All you can see is the surface around it beyond which light cannot escape—its event horizon. Astronomers are now convinced there are black holes all over the place, especially at the cores of all big galaxies—our own included.

So our universe is "shaped" like a Swiss cheese—full of tiny holes. However, this doesn't tell us whether the whole cheese is round or flat, cylindrical or toroidal. To work this out, we have to think a lot harder about the geometry of space-time. Before we can do that, though, we need to go back and think some more about laws of nature.

SYMMETRIC LAWS...

Despite the baffling complexity of the events that go on in our universe, there is an undercurrent of law and order into which all the hubbub fits. It's impossible for our brains to encompass every detail of what the universe is doing; in fact it's impossible for our brains to encompass every detail of what we are doing, even though our

brains are controlling our actions. There's just too much going on. But we've discovered a trick to make the complexities of the universe manageable. We formulate rules simple enough to be understood, yet accurate enough to give us real insight into what the universe is up to.

Scientists used to think that these rules were a description of how the universe actually worked, as the phrase "law of nature" attests. In Newton's day, his particular mathematical formulation of the law of gravity was thought to be an exact statement of how gravity acts. Thanks to Einstein, we now know that it's a very accurate approximation, which breaks down under extreme conditions. Even today, many physicists are convinced that the latest versions of nature's laws are true— they think that past attempts were approximate but what we have now is error free. They could be right, but history suggests otherwise.

Einstein had a particularly deep understanding of nature's underlying simplicities and, as I've said before (*see pp. 54–55*), he founded his view of physics on a symmetry principle. The symmetry transformations of space-time must leave the laws of nature unchanged. Relativity is the working out of this principle in the context of electromagnetism and gravitation—two key forces that govern how matter behaves.

To these two forces quantum theory has added two more—the strong and weak nuclear

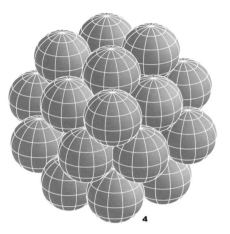

forces. And it has added a whole series of symmetry principles, which constrain the laws of quantum mechanics in the same way that the symmetries of space-time constrained relativity. Some of the symmetries involved are readily comprehensible: mirror-reflection; time-reversal (also referred to as time-reflection); and charge conjugation (parity), which interchanges positive and negative electric charge. Others find their expression only in the mathematics of the quantum world.

The core of quantum theory is particle physics, which seeks to enumerate and organize the tiniest building blocks of matter, the fundamental particles. Originally this looked easy, because only three such particles were recognized—protons, neutrons, and electrons. But physicists rapidly added more—photons, neutrinos, kaons, pions… Soon there were hundreds, all allegedly equally "fundamental," which wasn't a happy state of affairs.

In 1962 the American physicist Murray Gell-Mann and Israeli theoretical physicist Yuval Ne'eman discovered that a subclass of the fundamental particles, known as hadrons, possesses a beautiful internal symmetry. By transforming the mathematical equations that represent these particles according to some symmetries known as SU(3), for instance, we can in effect "rotate" a proton into a neutron.

That is, we can turn the equations for a proton into the equations for a neutron. Nature acquired a deep and exotic internal structure, in which even the identity of a particle was mutable.

The central objective of today's physics is a complete unification of all four forces of nature—often called a theory of everything. There are philosophical disputes about how useful or important such a theory would actually be—it is unclear, for instance, that it would add much to our understanding of psychology or economics, or even crystallography. Nonetheless, it would be a magnificent achievement to bring quantum theory and relativity under the same banner. On the theoretical front, there is a lot of excitement about so-called superstrings, which are like particles but are like curves instead of points. The entire theory of superstrings is driven by elegant mathematical symmetries, like those found by Gell-Mann and Ne'eman, but even more exotic.

Unfortunately, experimental evidence for superstrings is nonexistent and will be hard to get because the energies involved are far beyond the reach of today's apparatus. Nonetheless, the belief that the key laws of nature are expressions of deep symmetries of the universe remains at the heart of physics.

...FOR A SYMMETRIC UNIVERSE?

Curiously, the deepest symmetries of all do not describe our universe as it now is. They may describe it as it was shortly after the Big Bang, or they may be mathematical inventions that never actually applied to our universe at all. The source of this difficulty is the discovery that some of the apparent symmetries of our universe occasionally fail—they can be broken. This first became apparent in 1956 when the theoretical physicists Tsung-Dao Lee and Chen-Ning Yang suggested that the weak nuclear force violated mirror symmetry and the experimental physicist Chien-Shung Wu proved they were right. The laws obeyed by our universe are actually different from the laws for its mirror image. The violation of mirror symmetry seemed to be confined to the weak nuclear force (gravity), electromagnetism and the strong nuclear force

would work exactly the same way in a mirror world. Moreover, the degree of asymmetry of the weak force was relatively small. Everything made sense if our universe's laws were a small perturbation of the laws of a more perfect, more symmetrical universe, one in which all four forces were mirror symmetric.

Some of the most attractive mathematical models of the Big Bang predict something even more elegant—at one time all four forces were the same. But as the universe began to cool from the incandescent heat of its explosive birth, it underwent a series of phase transitions—like steam turning to water and then to ice. In a sense this is closer to the literal than the metaphorical, our universe's laws crystallized into a form in which the four laws split apart and each acquired its own distinctive properties.

Mathematically, the laws of that early universe were simpler, and more elegant, than

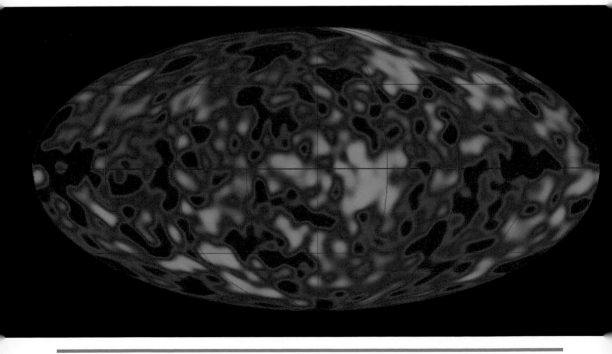

those of our current universe. It is unclear whether the ideal, all-too-perfect universe with a single force ever really existed, or whether it is a mathematical fiction that explains genuine almost-symmetries by pretending they once were exact. To continue the analogy with ice, it is logically conceivable that a universe made of ice might always have been made of ice, and that at no time in its past did liquid water ever exist. Nonetheless, we might gain mathematical insights into the structure of ice by representing its crystal lattice as the consequence of broken symmetry in a hypothetical universe of liquid water, or an even more symmetric universe of steam. Mathematical stories about the world can have a useful punch line even if the drama was never actually played out the way the mathematics suggests.

What we can be sure of is that in the very first instants of the Big Bang, our universe underwent

LEFT Ripples at the edge of time...unevenness in the cosmic background radiation, detected by the COBE satellite, shows that the early universe was slightly clumpy. Clumpy enough for gravity to create even more clumpiness, leading to the observed distribution of matter into clumps on every scale.

a different kind of broken symmetry. The distribution of matter, for example, was initially uniform but quickly started to become clumpy. From our point of view, this clumpiness was crucial to our own existence, because the clumps were seeds that caused matter to accumulate into galaxies, stars, and planets. In the early 1990s the COBE satellite gained headlines worldwide for detecting ripples at the edge of time—traces left today by that first change from uniformity to clumpiness. The degree of clumpiness was very small—about one part in ten thousand. Nevertheless, as the universe expanded and continued to cool, it was enough to lead to the fractal web of voids and superclusters that we find today.

Clumpiness notwithstanding, one of the other big puzzles of cosmology is why today's universe is so flat. If you take away the clumps and look at the background shape of the universe, it has remarkably little curvature. Gravitationally, the universe is like a flat desert dotted with hills and hummocks. The hills are the clumps of matter, but what they sit on is basically flat. To see that it could be otherwise, imagine that the desert is curved into a sphere or rolled up into a torus before the hills and hummocks are added.

Where does the flatness come from?

The favored theory is a scenario called inflation, in which the rate of expansion of the universe suddenly accelerated by an enormous amount before settling down to its current value. Think of a balloon being blown up into a small sphere and suddenly expand it a billionfold into a gigantic sphere. Any localized region of such a sphere would be indistinguishable from a flat plane—just as the surface of the Earth appears flat even though it is actually round. Cosmologists continue to argue about the precise details of the era of inflation, but they are convinced that something very like it must have happened to create the kind of universe we see today.

END OF THE WORLD

The flatness of today's universe is related to another big puzzle—how will the universe end?

According to general relativity, the more matter there is, the more curved the universe becomes. If there is a lot of matter, then the attractive force of gravity—which, on big scales, holds the universe together and slows the rate at which everything flies apart—will eventually win out, and the expanding universe will slow down, halt, and start to collapse. With less matter, the universe will expand forever. Because our own universe is so close to being flat, it seems to be poised right at the boundary between continual expansion and ultimate collapse.

Against this view we must set some dissenting evidence. We can survey the amount of matter in the regions of the universe that our telescopes can now see, which is getting pretty big nowadays, and we can estimate the total amount of matter in the universe. If what we can see is all there is in the region we're observing, then the universe contains only ten percent of the matter needed to reverse its expansion. This is difficult to believe since by the same token it also contains only ten percent of the matter needed to make it as flat as it is. Clearly there is a major discrepancy between what we observe and what our theories indicate.

Cosmologists have largely pinned their hopes on the presumed existence of the other 90 percent, in the form of cold dark matter—an exotic form of matter that cannot be observed with our current instruments and which could be spread throughout space, invisible to everything save the force of gravity. Critics object to this suggestion as a form of theory-saving, analogous to suggesting that the Moon is in fact inhabited, but by creatures that are invisible, inaudible, and intangible. Scarcely a month goes by without some announcement that the "missing mass" has been "found"—in gas between the stars, in quantum particles 3,000 light-years across with virtually no mass. None of this work is yet considered conclusive, and it is certainly possible that the problem lies not with missing mass but with our cosmological theories. After all, it's not an area of science where laboratory experiments can be used to settle difficult issues.

Supposing that the theories are right and the missing mass is there, we are probably just on the side of the critical mass that will eventually cause the expanding universe to slow down, stop, and collapse. From the Big Bang in the past, the universe will head inexorably toward a future Big Crunch.

This poses problems for the second law of thermodynamics, which is generally interpreted as an inescapable increase in the disorder—the technical term is "entropy"—of the universe. The problem is simple—the universe starts as a highly ordered system, a single point, and starts to expand. As it does so, its entropy increases. So far, so good. But now the universe stops expanding and starts to shrink—yet still its entropy must increase. Finally it returns to its original state, a single point, and the entropy must still be increasing—yet the final entropy

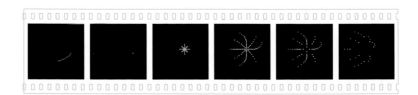

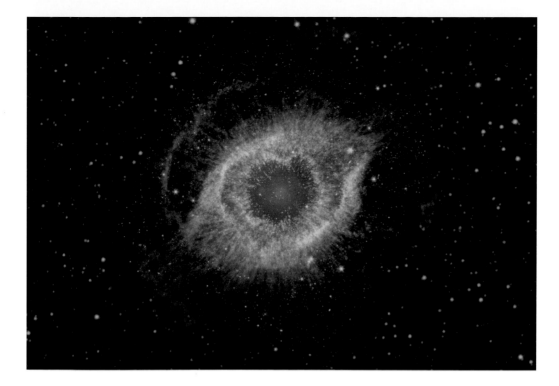

has to be the same as it was to start with, since the Big Crunch ends where the Big Bang started.

A lot of effort has gone into attempts to resolve this issue, and a lot of fascinating speculations have emerged. The direction of entropy increase appears to be related to the arrow of time, so some physicists think that when the expansion of the universe comes to a halt, so does time, and when the universe starts to contract, time begins—if that is the word—to flow backward. An increase of entropy in backward time is just like a decrease in forward time, so the apparent contradiction vanishes.

I think a more plausible resolution lies in the nature of the second law of thermodynamics and its realm of validity. It grew from work on steam

ABOVE AND BELOW How will the universe end? One possibility is that its expansion will gradually slow, and then reverse. The final minutes of the universe might be a Big Crunch: the Big Bang run backward.

engines and the impossibility of perpetual motion using thermal energy, and in the domain where it first arose, it works very well. However, as already discussed (*see pp. 170–171*), entropy increase is appropriate for systems like gases, with short-range repulsive forces acting; it is inappropriate for gravitating systems, with long-range attractive forces. Since the universe combines both types of force, entropic models will be appropriate for some questions but not for others, and gravitic models will be appropriate for some questions but not for others.

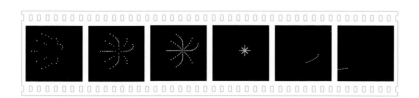

The increased "order" caused by gravitational clumping can offset the decreased "order" caused by thermal diffusion—the entropic books need not balance.

TIME TRAVEL

Modern cosmology has revived an old dream about time—not its reversal, but the ability to travel along it. In 1894/95 the *New Review* serialized H. G. Wells's famous science fiction story *The Time Machine*, in which the Time Traveler uses advanced technology to explore the future of the Earth and finds that humanity has speciated into the innocent but placid Eloi and the savage Morlocks. Ever since, time travel has been a standard trope of science fiction, along with all of its strange paradoxes. There is the Grandfather Paradox—what would happen if someone went back in time and killed her own grandfather? Well, she wouldn't have been born, so she couldn't have gone back and killed him, so she would have been born, so… And there is the Cumulative Audience Paradox—major historical events would attract time-traveling tourists from the indefinitely far future. So, for example, the Battle of Hastings would have been surrounded by millions of spectators hoping to catch the death of King Harold. But we know, from historical records, that no such crowd was present.

All of this might have remained mere fictional speculation had a generation of physicists who had grown up with science fiction not begun to wonder whether the laws of nature—as currently conceived—forbid time travel or permit it. And they found, perhaps to their surprise, that it is permitted.

Nothing in the current formulation of the laws of physics says that time travel is impossible.

There are two ways to interpret such a discovery. One is that despite all this, time travel is evidently impossible (because of the causal paradoxes) so the current laws of nature need adjusting to incorporate such an obvious gap. The other is that time travel is possible, in which case we have to explain why the paradoxes aren't actually paradoxical. An early scenario for time

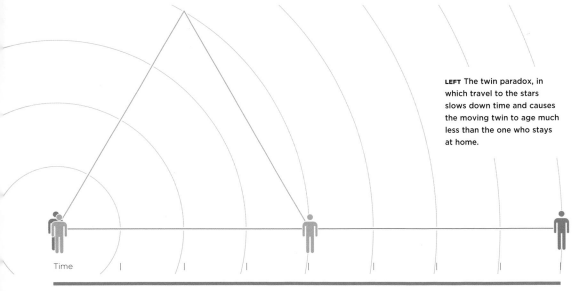

LEFT The twin paradox, in which travel to the stars slows down time and causes the moving twin to age much less than the one who stays at home.

Time

travel exploited three features of cosmology. One is the relativistic intertwining of space and time; the second is the existence of black holes; the third is the time-reversibility of the laws of nature. The mathematics of relativity leads to the so-called Twin Paradox. Tweedledum and Tweedledee are identical twins, but one day Tweedledum heads off to a distant star at close to the speed of light, gets there, turns around and comes home again—also at close to the speed of light. According to relativity, the passage of time is slower for Tweedledum than it is for stay-at-home Tweedledee, so when the voyager gets back, he is several years younger than his previously identical brother.

Next ingredient—a black hole, which sucks matter in but won't let it escape. Final ingredient—time-reversibility. If a black hole can exist, so can its time-reversal: a white hole, which spits matter out but won't let it back in. Join the two end to end and you have a wormhole—a short-cut through space which sucks matter in at one end and spits it out again at the other. Thanks to the Twin Paradox, a short-cut through space can now be turned into a short-cut through time. Replace Tweedledum by the black hole end of the wormhole, and let Tweedledee take care of the white hole end. Then time passes more slowly at the white hole end than it does at the black hole end. If you travel out to the black hole end and come home by taking the short cut through the wormhole, you move several years into the past. More precisely, you traverse a closed time-like curve whose future runs into its past. This is the cosmologist's formalization of "time machine."

There are problems with a wormhole as a practical time machine—if you try to shove a massive object through it, such as a human being, then the hole snaps shut before there is time to get through. One solution to this problem is to thread "exotic" negative-energy matter through the wormhole to hold it open. The snag is no such matter is known to exist, although quantum-mechanically it ought to be possible.

Another solution is to use not a wormhole but cosmic string—threadlike gravitational singularities that create curvature of space-time along a curve.

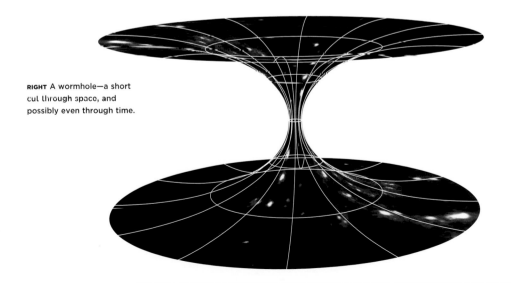

RIGHT A wormhole—a short cut through space, and possibly even through time.

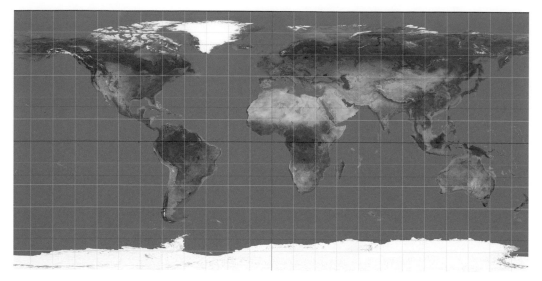

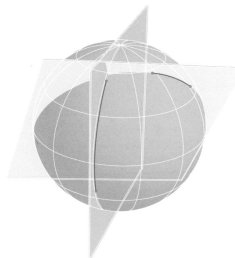

If two cosmic strings pass close to each other, traveling in opposite directions at nearly the speed of light, once again there exists a closed timelike curve. The snag this time is that there isn't enough energy in the universe to build such a piece of machinery. So the desktop time machine must, for now, remain a science fiction writer's dream.

NON-EUCLIDEAN GEOMETRY

We're now ready to think again about the shape of the universe. New physics demands new geometries. To most people, the word "geometry" means the geometry of straight lines and circles in the plane—the concept of geometry that goes back to Euclid, an ancient Greek who lived around 300 BCE. But even then it was becoming clear that other kinds of geometry might also be important for understanding the world around us. Indeed, "around" is the word, for the Greeks recognized that the Earth was (roughly) spherical in shape.

The need for ships to navigate accurately across the face of a spherical Earth led to the development of a second kind of geometry—

ABOVE AND BELOW Our round Earth (below) is large enough that human-sized regions look almost flat, so it would be easy to confuse its curved geometry with the flat geometry of a plane. Because the Earth is not flat, any flat map of it must distort the shapes of the continents (top left). Careful compromises can produce flat maps that come a lot closer to the true shapes, but only by tearing the globe along parts of its oceans (above right).

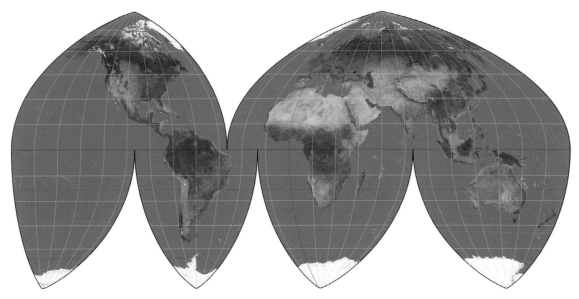

spherical geometry. On a plane, a geodesic—the shortest path between two points—is a straight line. On a sphere, however, geodesics are circles formed by cutting the sphere with a plane that passes through the sphere's center. Examples include all lines of longitude, but not lines of latitude (other than the equator), which are circles lying in planes that do not pass through the center of the Earth. As a result, spherical geometry differs from Euclidean geometry in many ways. Perhaps the most dramatic of these is the sum of the angles of a triangle. In Euclidean geometry, this is always 180 degrees. In spherical geometry, however, the sum of the angles is always more than 180 degrees, and the excess is proportional to the area of the triangle. This difference in geometrical properties causes endless problems when trying to make a map of a round Earth on a flat sheet of paper. It is not possible to represent geodesics on the Earth as geodesics on the map and to keep angles correct. So maps are designed to preserve some features of the geometry, but not others. The choice of which features to preserve leads to innumerable distinct map projections—methods for

representing a round world on a flat map. The best known is the Mercator projection, introduced in 1569, which maintains compass courses (directions) but distorts areas, making regions near the poles seem far larger than they really are and countries proportionally incorrect.

When Euclid wrote the *Elements*, he laid down a series of fundamental assumptions, or axioms. One of them seemed far more complicated than the rest. It asserted that given a line and a point not lying on that line, there exists a unique parallel line passing through that point. People wondered whether it could be deduced from the other axioms.

Spherical geometry nearly obeys all of Euclid's axioms, except the axiom of parallels. On a sphere any two lines meet—there are no parallels. However, one of Euclid's other axioms says that two lines meet in a point, while two lines on a sphere always meet in two (diametrically opposite) points. Modern mathematicians happily get around this by redefining "point" to mean a pair of diametrically opposite points, but in those days a point was a point and two points weren't.

Out of the attempts to find a proof of the parallel axiom came a different alternative to Euclid's geometry, one in which parallels exist but are not unique. One way to visualize this new geometry, known as hyperbolic geometry, was invented by Henri Poincaré. Fix a circular disk in the plane, and insist that the only points that exist are those interior of (and not on the rim of) this disk. Define a line to be a circle that cuts the rim at right angles—or, more precisely, that part of such a circle that lies inside the disk. Then, miraculously, all of Euclid's axioms are satisfied for this geometry, except the parallel axiom. If the parallel axiom could be proved using only the other axioms, then it would have to be valid in hyperbolic geometry—but it isn't.

Another way to interpret hyperbolic geometry is as the geometry of geodesics on a surface of constant negative curvature. Spherical geometry is the geometry of geodesics on a surface of constant positive curvature, and Euclidean geometry is the geometry of geodesics on a surface of zero curvature—the plane.

The unifying concept, then, is curvature.

CIRCLES IN THE SKY

The Earth is a curved surface—a sphere.

What shape is the universe? Is it curved? If so, how?

Until fairly recently it was thought that the universe was infinite, modeled by three-dimensional Euclidean space, but we are now confident that this is wrong. Indeed we think that the universe is finite, though without any edges where space would simply "stop" like hitting a wall.

In two dimensions, the surface of a sphere is finite but has no edges. Nevertheless, creatures that were limited to moving on such a surface, and were unaware that anything else existed, would be able to deduce its shape by observing

its geometrical rules. We hope to do the same for our own universe.

We've seen that Euclid's geometry is not the only possible geometry. As well as Euclidean geometry—the flat plane—there can be positively curved space, in which the angles of a triangle add up to more than 180 degrees. And there can be hyperbolic geometry, negatively curved space, where the angles of a triangle add up to less than 180 degrees. The German mathematician Georg Bernhard Riemann discovered that similar ideas apply to spaces of three or more dimensions. Perhaps our own universe is actually curved—meaning that really big triangles do not behave like small Euclidean ones, not that the universe is wrapped around something else. Albert Einstein suggested that it is the curvature of space that causes gravity.

We currently think that the universe began with the Big Bang, when space and time both came into existence as a single point that rapidly grew. What shape, though, did the universe grow into? The obvious possibility is that it is a 3-sphere, the three-dimensional analog of the surface of a sphere—positively curved, finite, without edges. This is the shape assumed in early solutions of Einstein's equations. But there is another intriguing possibility: that the universe is negatively curved—and still finite. There are many such geometries—indeed, infinitely many. Mathematicians call them hyperbolic manifolds.

They can be constructed by creating tiling patterns in Poincaré's disk. Hyperbolic geometry is unusually well equipped with tilings. For instance there are innumerable different triangles that can be used to tile the Poincaré disk in symmetric ways. In Euclidean geometry there are fewer symmetric tilings, the simplest being a grid of squares. Suppose you exercise some mathematical imagination and pretend that all squares in such a grid are the same, so

that when a point moves off the edge of a square it returns to the opposite edge, "wrapping around" like the screen in many computer games. What shape is this kind of tiled universe? It is a torus, the surface of a doughnut with a hole. If you glue opposite edges of a square together, which is what this construction in effect does, then gluing the first pair of edges rolls the square into a tube, and gluing the second pair bends the tube around full circle and joins its ends. That's a torus. A torus is finite and has no boundary.

Hyperbolic manifolds arise when you play the same game with tilings of the Poincaré disk. Start with a symmetric tiling, and pretend that certain tiles that look distinct are really the same. This "wraps up" the tiling to create a shape that still has constant negative curvature, but is finite and has no boundary, just like a torus. The suggestion is that our own universe is one of these exotic shapes, but in four dimensions of

space-time rather than the two of Poincaré's disk. If so, how could we tell? In any finite universe without edges it should be possible to observe the same galaxy in more than one position in the sky, by looking around the universe in different directions. Our universe is gigantic, but today's telescopes are becoming very powerful. Mathematicians have worked out that if the universe is negatively curved, there will be special circles in the sky where the same patterns of distant galaxies repeat identically. Observe these circles, and you can work out what shape the universe is. To find them, we need to use a computer to compare the galaxies on all possible circles. This is a huge task, but telescopes are being readied to survey the sky and computers are being programed to search for duplicated circles.

As awe inspiring as it may sound, the shape of the entire universe may soon be within our grasp.

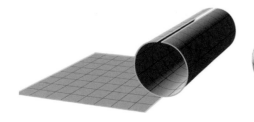

16

THE ANSWER

It's amazing how far a simple question about an everyday event can lead. We started out wondering about the shape of a snowflake, and we've worked our way up to deep philosophical questions about the foundations of physical law, the nature of space, time, and matter, and the shape and history of the universe. We've encountered entirely new kinds of geometry, and we've expanded our intellectual horizons to encompass not just a tiny sixfold-symmetric speck of crystalline ice but also all of the marvelous patterns of the physical world—and the biological patterns, too, for many of nature's most delightful and intriguing regularities are to be found in living creatures.

In fact, living creatures are alive because of their constituent patterns. However, unlike ordinary nonliving matter, life has evolved the ability to harness these patterns—to make sure they fit together in specific particular ways, so that the strange self-referential processes that we associate with life can work reliably. So that life can perform miracles—it can reproduce itself. Life can complicate itself. Life can organize itself. These are not processes that we expect to find in ordinary, inorganic matter. Indeed, if we did find them, we would immediately conclude that we were looking at organic matter. If it reproduces like a duck and self-organizes like a duck, then it's alive like a duck and it will quack like a duck.

And yet this is at heart a ridiculous distinction. There is no such thing as organic matter. A live duck is made from the same kinds of atoms as a dead duck, and they all obey the same laws as the atoms in a rock, an ocean, or the eleventh moon of Saturn. Organic matter is simply matter organized in a certain way—it is the system, not the components, that possesses

the remarkable self-organizing properties. The flexibility and adaptability of life, paradoxically, has arisen from the inflexibility and rigidity of natural law.

Intuition is taking a real pounding here. As we have seen, complexity and simplicity do not pass unchanged from rules to consequences. Symmetry and continuity do not pass unchanged from rules to consequences. And now rigidity and inflexibility do not pass unchanged from rules to consequences. Indeed, it's not clear that anything meaningful passes, unchanged, from rules to consequences.

Except—potential.

The consequences unfold the potentialities inherent in the laws and, crucially, in the context in which those laws operate. The laws for a carbon atom are the same whether it is inside a duck or inside the eleventh moon of Saturn—but the role that it plays and the way it carries out that role is very different in these two cases.

What does all this tell us about snowflakes?

For a start, it tells us that snowflakes ought to be comprehensible. If we can hope to discover

the shape of the universe, then we ought to be able to handle a snowflake.

Conversely, it tells us that whatever explanation we give for the shape of a snowflake, it won't be the last word on the matter. The best we can do is to tell convincing stories about snowflakes. Good enough to let us see that they make sense, good enough to suggest interesting experiments that actually work, good enough to make snowflakes to order, in the laboratory. Beyond that, humans cannot and should not aspire. The Truth (with a capital "T") must lie forever outside the realms of our understanding (if indeed there is such a thing as The Truth, which I very much doubt). What we can aspire to is truths, with a small "t," that is to say, scientific stories that work in their own limited realm— and work surprisingly well given the simplicity of their ingredients.

Are snowflakes cast in some meteorological mold? *No.*

Are they grown according to some unknown cosmic recipe? *Well, not exactly.*

ABOVE Why do snowflakes have such amazing shapes? By putting together all we have discovered, we can give a satisfactory answer. Although, of course, in science every new answer leads to new questions.

Does their form emerge from the laws of physics by a process too complex to grasp in complete detail? *Certainly.*

Can we describe that process schematically, well enough to gain useful insights? *Undoubtedly.*

Is the process a phase transition? *Yes.*

Is it a bifurcation? *Yes.*

Is it symmetry-breaking? *Yes.*

Is it chaos? *Yes.*

Is it a fractal? *Yes*

Is it a complex system? *Yes.*

Will we ever understand a snowflake completely? *No.* But let's see how far we can get.

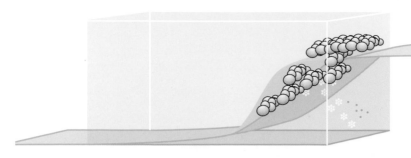

LET CHAOS STORM

Snow is born in clouds.

Seen from space, the Earth is mainly blue and white—oceans and clouds. The greens and browns of the continents are also often visible, but are often obscured by clouds.

Seen from up close, clouds occur in an almost unlimited variety of forms. A cloudy sky changes from one moment to the next, seemingly never repeating its patterns. Clouds are mainly composed of water vapor, condensing from the atmosphere as warm air from lower regions rises to cooler heights. Because it is to some extent the movement of air that creates clouds, the atmosphere within many clouds is quite dynamic. We have already seen that there is order to this constantly changing variety—the famous cumulus, nimbus, cirrus, and stratus, and their more modern refinements. We now have a good idea how these forms relate to the underlying physics.

Clouds in the lower levels of the atmosphere interact strongly with the ground (or ocean) beneath. If the ground is hot, then it warms the lower layers of air. Hot air rises, but because the entire layer of air cannot rise uniformly, the symmetry breaks. This creates localized convection cells, in which air rises at the center and falls, having cooled, at the edges. The rising air carries moisture with it, and if it is humid at ground level, the quantity of moisture is especially great. As the air rises, it cools, and as it cools, the moisture precipitates out as liquid droplets of water or as ice—a phase transition.

The weather maps that we see on our television screens attempt to summarize a lot of information about the state of the atmosphere in terms of a few simple concepts—wind speed and direction, temperature, bands of rain or hail or fog—and fronts. A front occurs when a mass of air moves into a region occupied by air that is significantly warmer or colder. The concept of a front was introduced during World War I, and in state-of-the-art meteorological research it has pretty much been replaced by more sophisticated concepts, but it provides a convenient and familiar image with which to discuss the formation of snow.

A warm front occurs when a mass of warm air invades a region currently occupied by colder air.

The warm air is less dense, so it rides above the cooler air. Sandwiched between is a wedge-shaped mixing zone, where turbulent flow stirs the two masses of air together. Above the mixing zone, a thick layer of nimbostratus cloud is formed as the warm, moisture-laden air is forced upward and cools. This is the heavy gray cloud layer that we usually associate with rain. Higher still, and farther ahead of the front, are altostratus and altocumulus clouds. Below them, in the remaining mass of cool air, clumps of stratocumulus may occur.

Within the mixing zone, in regions where the temperature change is most abrupt, excess water vapor condenses to form ice crystals, which can either grow into flakes of snow or dense lumps of hail. They circulate within the cloud and eventually descend below its base. Depending on the temperatures lower down, the falling ice either melts to produce rain or remains frozen to deposit hail or snow. Most of the rain falls ahead of where the front meets the ground.

A cold front is similar, but the incoming air drives itself under the warmer air like a wedge. Nimbostratus clouds appear in a band, slightly ahead of the front; above and slightly behind these, the cloud rises and becomes altostratus. In the lower layers, fine-weather cumulus clouds appear behind the advancing front, while stratocumulus clumps gather ahead of it. Rain, hail, or snow fall from the base of the nimbostratus cloud, roughly where the front meets the ground.

Snow can also form in cirrostratus clouds, high up in the atmosphere. Snow formed in this manner can fall from a cloudless sky if warm air overlays cold air, which is a common phenomenon in the polar regions. Where the warm and cold air meet, the cold air can become supersaturated with water vapor. Then the vapor precipitates as small needles and columns of ice and drifts downward as glittering "diamond dust."

LET SNOWFLAKES SWARM

Experimentalists can grow ice crystals in apparatus that ensures much simpler conditions than those that prevail in a storm cloud—constant temperature, constant pressure, and so on. It then becomes possible to find out what

factors influence the shape of the crystal. It turns out that the two main factors are temperature and supersaturation.

Temperature is a familiar concept, and we all know that it has to be cold for ice to form. Supersaturation refers to the amount of water vapor present in the air. Normally, a given volume of air can hold a limited amount of water vapor. Above this saturation level, the excess vapor condenses out as a fine mist. Warm air has a higher saturation level than cold air, that is, it can hold more water vapor. If saturated warm air is cooled sufficiently smoothly, it can become supersaturated, containing more water vapor than the normal saturation level at the lower temperature. This state is metastable, meaning that sudden disturbances, dust particles, or other irregularities can trigger a change resulting in the correct saturation level. The excess water vapor is then expelled and if the temperature is low enough it forms ice rather than mist. All of these changes from water vapor to liquid water or ice are phase transitions in a system of water molecules.

Ice crystals form a variety of shapes. The simplest, hexagonal plates, form at temperatures just below freezing—between 32°F (0°C) and 27°F (-3°C)—and low levels of supersaturation (less than 30 percent). The reason is that at these temperatures a straight edge of an ice crystal grows stably—any small irregularities get filled in and the edge remains straight as it grows. Taking the sixfold symmetry of the crystal lattice of ice into account, we see that the growing crystal makes six copies of a straight edge—a hexagon. And as we have already

discovered, ice crystals prefer to grow in flat layers, at least in conditions that are not too extreme, so we find flat hexagonal plates (*see pp. 114–115*).

At the same temperatures, but at higher levels of supersaturation (above 30 percent) a phenomenon known as the Mullins-Sekerka instability sets in. The translational symmetry of a flat edge breaks and the dynamics undergoes a bifurcation. Now tiny irregularities become amplified and the growing edge acquires spikes. The edges of the spikes cannot remain flat if the spikes get too long, so secondary spikes split off sideways. This growth process is similar to tip splitting in fractal growth processes (*see pp. 168–169*), and it leads to a fractal shape for the crystal. The geometric regularity of the crystal lattice makes this fractal growth process generate fernlike dendritic crystals.

Around the 30 percent supersaturation level, we find many other forms of ice crystal, depending on the temperature. Between 32°F (0°C) and 27°F (-3°C), as we've just seen, the crystal is dendritic. Between 27°F (-3°C) and 23°F (-5°C) the crystals are needle-shaped. Between 23°F (-5°C) and 18°F (-8°C) the thickness of the crystal increases and the ice forms hollow hexagonal prisms. Between 18°F (-8°C) and 10°F (-12°C), and again between 3°F (-16°C) and -11°F (-24°C), we observe sector plates—thin plates with symmetric decorations. Between 10°F (-12°C) and 3°F (-16°C) dendritic crystals reappear, and below -11°F (-24°C) we find hollow prisms. At lower supersaturation levels, if the temperature is low enough, the crystals tend to be thicker—we find thick, solid plates and solid prisms.

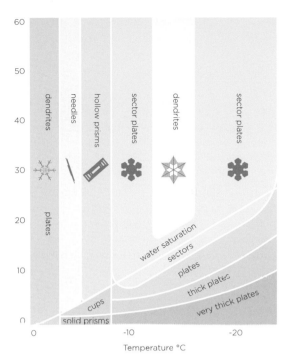

LEFT Different atmospheric conditions lead to different shaped ice crystals. The most important factors are temperature and supersaturation—available moisture. The numerical values of these factors determine the general shape of the crystal. Fine details depend on the chaotic conditions in the clouds.

These are by no means the only forms, but the important point here is the diversity of hexagonally symmetric shapes and their sensitive dependence on the atmospheric conditions inside the cloud at the place where the crystal is forming. The physics of ice crystals provides a wide variety of forms, and the precise shape of any given snowflake depends on the exact sequence of forms through which it passes as it makes its erratic journey through the cloud, accreting water molecules as it does so.

The combination of regularity (symmetry) and irregularity (chaotic dynamics in the cloud) solves our puzzle of how nature manages to make snowflakes that have sixfold symmetry but are otherwise enormously diverse. A lot of deep things go into the shape of a snowflake— phase transitions, symmetry-breaking, bifurcation, fractal geometry, chaos. Snowflakes are a showcase for the mathematics of pattern formation.

I WAIT FOR FORM

What shape is a snowflake?

Any shape that it wants to be—but snowflakes have no wants, and while they are clothing themselves in physical form they are not yet snowflakes. They come into being in huge, towering clouds of water vapor—atoms of hydrogen and oxygen, bound together in threes, scarcely noticing their neighbors until they crash into them. The dance of the water molecules is pure physics, complex but disorganized, mass movement of infinitesimal molecules. The dance has statistical patterns, which we call temperature, pressure, saturation. These patterns set the stage for the molecular dance, and particular combinations of those patterns change the dance's rhythm and tempo. Previously dissociated molecules bind together into a diminutive crystalline seed, no longer vapor but solid. The mathematical regularities of matter and the forces that bind it create tiny jewels of ice, its molecules clicking together with almost perfect precision. The structure that they make betrays its reliance on universal rules, for it is a hexagon.

The storm cloud is a complex system of molecules, seething with billions of particles of hexagonal dust. Turbulent convection currents whisk them skyward, buffet them this way and that, plunge them toward the ground far beneath. More molecules collide with them and stick where they hit, clicking into place in the growing lattice. Again the rules reveal their hidden influence, for out of the disorder patterns emerge. The visual form of these patterns is potentially diverse, but again the diversity is pruned by statistical regularities of supersaturation and temperature. At each instant all six extremities of the incipient snowflake are exposed to almost identical

conditions, for the flake is far smaller than the scale on which the statistical properties of the cloud vary significantly. At each extremity the molecular rules build much the same structures—each flake retains its initial sixfold symmetry. From moment to moment, the form changes according to the flake's environment, so the symmetry encompasses elegant filigreed decoration. The movement of air and vapor in the cloud is chaotic, differing from one place to another, from one moment to the next. Each flake follows its own trajectory, experiences its own history, and also constructs its own tiny crystalline record of its journey through the storm…

In six identical copies. A billion hexagonal seeds, a billion journeys—a billion histories. A billion snowflakes. All repeating their sixfold patterns, and every pattern different. These are Kepler's six-cornered snowflakes. There are

others. The laws of physics tell many stories—this is but one.

The flakes drift erratically toward the bottom of the cloud. When the moment is ripe, they fall. Their fluffy platelets cover ground, bushes, trees—the world becomes frosted.

The snowflake itself is not the most astonishing thing here. The big surprise is that our universe is rich enough to produce not just a few complex forms, but complexity in such quantity that our entire planet is just one insignificant part of it. There is more complexity in a star than in a snowflake, and there are more stars in the universe than flakes in a snowstorm.

I am a mathematician. I experience these wonders through a mind that has spent a lifetime learning how to detect patterns, how to understand patterns, how to analyze patterns, how to use patterns, how to find new patterns… I stand on the shoulders (and lean on the elbows) of giants, on five thousand years of mathematical history that has been groping toward such understanding. I see what all humans see, and in a few respects perhaps I see more. I see clues to rules, laws, regularities. Where the child saw a fern on an icy window, the adult now sees the fractal growth of crystalline molecules and the hidden symmetry of nature's forces.

I do not believe that the universe is diminished through understanding, that the beauty of a snowflake can be spoiled by an awareness of what makes it. The universe is not a conjuror's magic, ruined if you know the trick. But more than all this, I'm aware of how little we truly know about our world, of how impoverished my story of the snowflake has to be in comparison to the glittering, frozen reality. There is so much more to learn.

What shape is a snowflake?

Snowflake-shaped.

GLOSSARY

Attractor

The state of a dynamical system changes with time. One way to visualize these changes is to plot the relevant variables—the quantities that characterize the system—in a diagram. As time passes, the state of the system moves along some path in this diagram. These paths often "home in" on some regions of the diagram to form a specific shape, called an attractor. It is a geometric description of the long-term behavior of the system.

Bifurcation, catastrophe

Sometimes a very small change in a system can lead to a big change in behavior. For instance the state may suddenly jump to a totally different one—say, a branch that is being bent suddenly snaps. Such a sudden change in state is called a bifurcation or a catastrophe.

Big Bang, Big Crunch

Astronomical observations suggest that the universe suddenly came into being about 12–15 billion years ago as a tiny speck of space-time that rapidly expanded. This is the Big Bang theory of the origin of the universe. The universe may end with its reverse, the Big Crunch, or it may expand forever.

Black hole, event horizon, wormhole

If a massive star collapses, its gravitational field can become so strong that light cannot escape. It is then a black hole. The threshold where light becomes trapped is its event horizon. A black hole can be joined to its opposite, a white hole, to form a wormhole—a shortcut through space-time.

Cellular automaton

A cellular automaton (plural "automata") is a mathematical system formed by an array of "cells"—say, squares like those on a chessboard. Each cell can be in one of a number of states, represented by colors. Specific rules determine how the color of each cell changes from one instant to the next: these rules depend on the colors of its neighbors.

Chaos

A system that obeys precise mathematical rules, with no explicit random element, can nevertheless behave in a surprisingly complicated way. In fact, some aspects of its behavior may appear to be random. Such a system is said to exhibit deterministic chaos, or to be "chaotic." Weather is an example.

Determinism

Isaac Newton and his contemporaries discovered that the physical universe can be described by mathematical equations. Those equations predict that the state of the system at some instant of time leads to only one possible future. This inspired the philosophy of determinism: in principle, the future of the universe is completely determined by its present.

Diffraction pattern

When X-rays pass through a crystal they interfere with each other to produce a pattern that is mathematically related to the crystal's atomic structure, called a diffraction pattern. From this, the structure of the crystal itself can be calculated.

Equitempered musical scale

In "natural" musical scales, such as those employed by humans when singing, the intervals between consecutive notes may differ slightly. The equitempered scale changes the pitch of the notes so that every interval is exactly the same. Its greatest early exponent was Johann Sebastian Bach.

Eukaryote, prokaryote

All life-forms on Earth (not counting viruses) fall into two major categories. Prokaryotes form a single, rather primitive "cell" with no nucleus or cell wall: the main examples are bacteria. Eukaryotes keep most of their genetic material inside a nucleus, and have a cell membrane. They may be single-celled (amoeba) or multicelled (grass, snail, pig, human…).

Fractal

A geometric shape that exhibits intricate detail, no matter how much it is magnified.

Hyperbolic manifold

A multidimensional curved space whose geometry in small regions is non-Euclidean. More specifically, small regions are negatively curved, so that many different lines, all passing through the same point, can be "parallel" to a given one.

Isomorphism

Two mathematical structures are isomorphic if they have the same abstract structure but are described in apparently different terms. Thus "one, two, three…" and "un, deux, trois…" use different words (in English and French) to describe essentially the same thing. Tiling patterns are isomorphic if the tiles fit together in the same way, and locally isomorphic if any finite region of one can be found somewhere in the other.

Mandelbrot set

This was a famous fractal invented by the Polish mathematician Denoit Mandelbrot. Its definition uses complex numbers $z = x+iy$ where $i = \sqrt{-1}$. Given any complex number z, form the sequence $0, z, z2+z, (z2+z)2+z,…$, where each number is the square of the previous one plus z. The number z belongs to the Mandelbrot set if this sequence does not escape to infinity.

Möbius band

A single-sided surface invented by the German mathematician August Möbius in 1858. Take a strip of paper, about 10in (25cm) long by 1in (2.5cm) wide. Join it end to end, but giving the strip a half-twist so that opposite sides join together. Now it has only one side: if you start painting one side red and keep going, you cover both sides of the paper, not just one.

Periodic cycle, oscillation

A cycle is a series of events that keep occurring in the same order. If the events take the same time to occur on each repetition, then the cycle is periodic. For example, the pendulum of a clock repeats the same swinging motion in the same time, over and over again. This kind of behavior is also called an oscillation.

Quantum mechanics/theory

From 1887 onwards physicists came to realize that on very small scales of space and time the laws of physics are quite different from those observed on a human scale. Particles can sometimes behave like waves, and energy comes in multiples of a fixed tiny unit, a "quantum." The resulting theory, quantum mechanics, now forms the basis of physics on the microscopic level.

Relativity, Twin Paradox

Famously discovered by Albert Einstein in 1905, relativity is the physics of space, time, and gravitation on very large scales and at very high speeds. Here, too, the laws of physics differ from those observed on a human scale. In Special Relativity, space shrinks and time slows down as objects approach the speed of light. One consequence is the Twin Paradox: if one twin travels to a distant star and home again, very rapidly, they are younger than the other twin when they get back. In General Relativity, gravity is caused by the curvature of space-time.

Supersymmetry, superstring, Theory of Everything

Physicists hope to unite relativity and quantum mechanics in a single theory that will underpin all of physics. Any theory of this kind is known as a Theory of Everything. One candidate is the theory of superstrings, which replaces fundamental particles by closed loops that vibrate. The theory is based on the phenomenon of supersymmetry, in which the laws of quantum mechanics remain the same when certain particles are replaced by hypothetical particles that are their mathematical transforms.

Symmetry

A symmetry is a mathematical transformation that leaves some object unchanged. In bilateral (also called "left–right" or "mirror") symmetry, the object looks the same as its reflection. In rotational (also "radial") symmetry, the object can be rotated through various angles. In dilational symmetry, the object can be magnified or shrunk.

Thermodynamics, second law

Thermodynamics is the theory of heat, temperature, pressure, and other such quantities in gases, and it is based on a model in which the atoms of the gas are small spheres that bounce off each other. Physicists recognize several laws of thermodynamics, of which the most famous is the second. This says that associated with any thermodynamic system there is a quantity called entropy, usually interpreted as "disorder," which always increases as time passes.

Vortex

A vortex (plural "vortices") is a region of a fluid that is spinning. Vortices can be large (Jupiter's red spot) or small (a smoke ring). They can occur in liquids (a whirlpool) or gases (a hurricane). Invisible vortices in air can sometimes generate lift, enough to support a bee—or a jet airliner.

FURTHER READING

Art

Abas, Syed Jan and Amer Shaker Salman, *Symmetries of Islamic Geometrical Patterns* (World Scientific, 1995).

Critchlow, Keith, *Islamic Patterns* (Shocken, 1976).

Field, Michael J. and Martin Golubitsky, *Symmetry in Chaos* (Oxford University Press, 1992).

Schattschneider, Doris, *Visions of Symmetry: Notebooks, Periodic Drawings and Related Work of M. C. Escher* (Freeman, 1992).

Bifurcation and catastrophe

Poston, Tim and Ian Stewart, *Catastrophe Theory and Its Applications* (Pitman, 1978).

Zeeman, E. C., *Catastrophe Theory: Selected Papers 1972–77* (Addison-Wesley, 1977).

Biology

Gambaryan, P. P., *How Mammals Run* (Wiley, 1974).

Goodwin, Brian, *How the Leopard Changed its Spots* (Weidenfeld & Nicolson, 1994).

Gray, James, *Animal Locomotion* (Weidenfeld & Nicolson, 1968).

Watson, James, *The Double Helix* (Signet, 1968).

Cellular automata

Berlekamp, Elwyn R., John H. Conway, and Richard K. Guy, *Winning Ways* (Academic Press, 1982).

Gale, David, *Tracking the Automatic Ant* (Springer-Verlag, 1998).

Chaos

Gleick, James, *Chaos* (Viking, 1987).

Hall, Nina (ed.), *The New Scientist Guide to Chaos* (Penguin, 1991).

Ruelle, David, *Chance and Chaos* (Princeton University Press, 1991).

Stewart, Ian, *Does God Play Dice?* (Penguin, 1997).

Complexity

Casti, John, *Complexification* (Abacus, 1994).

Lewin, Roger, *Complexity* (Macmillan, 1992).

Mainzer, Klaus, *Thinking in Complexity* (Springer-Verlag, 1994).

Waldrop, Mitchell, *Complexity* (Simon & Schuster, 1992).

Fractals

Barnsley, Michael, *Fractals Everywhere* (Academic Press, 1988).

Mandelbrot, Benoit, *The Fractal Geometry of Nature* (Freeman, 1982).

Peitgen, Heinz-Otto, Hartmut Jürgens, and Dietmar Saupe, *Chaos and Fractals* (Springer-Verlag, 1992).

Geometry

Gray, Jeremy, *Ideas of Space* (Oxford University Press, 1979).

Greenberg, Marvin Jay, *Euclidean and non-Euclidean Geometries* (Freeman, 1993).

History and biography

Fauvel, John, Raymond Flood, and Robin Wilson, *Möbius and his Band* (Oxford University Press, 1993).

Gleick, James, *Genius: Richard Feynman and Modern Physics* (Little, Brown, 1992).

Kline, Morris, *Mathematical Thought from Ancient to Modern Times* (Oxford University Press, 1972).

Kragh, Helge S., *Dirac: A Scientific Biography* (Cambridge University Press, 1990).

Westfall, Richard S., *Never at Rest: A Biography of Isaac Newton* (Cambridge University Press, 1980).

Mathematics and nature

Meinhardt, Hans, *The Algorithmic Beauty of Sea Shells* (Springer-Verlag, 1995).

Prusinkiewicz, Przemyslaw and Aristid Lindenmayer, *The Algorithmic Beauty of Plants* (Springer-Verlag, 1990).

Stewart, Ian, *Nature's Numbers* (Weidenfeld & Nicolson, 1995).

Stewart, Ian, *Life's Other Secret* (Wiley, 1998).

Stewart, Ian, *Mathematics of Life* (Profile, 2011).

Stewart, Ian and Martin Golubitsky, *Fearful Symmetry* (Penguin, 1993).

Thompson, D'Arcy Wentworth, *On Growth and Form* (Cambridge University Press, 1942).

Philosophy

Barrow, John, *Theories of Everything* (Oxford University Press, 1991).

Casti, John L., *Paradigms Lost* (Scribners, 1989).

Casti, John L., *Searching for Certainty: What Scientists Can Learn about the Future* (Morrow, 1990).

Cohen, Jack and Ian Stewart, *The Collapse of Chaos* (Viking, 1994).

Davies, Paul, *The Mind of God* (Simon & Schuster, 1992).

Dyson, Freeman, *Disturbing the Universe* (Basic Books, 1979).

Dyson, Freeman, *Infinite in All Directions* (Basic Books, 1988).

Kauffman, Stuart A., *At Home in the Universe* (Viking, 1995).

Stenger, Victor, *The Fallacy of Fine-Tuning* (Prometheus, 2011).

Stewart, Ian and Jack Cohen, *Figments of Reality* (Cambridge University Press, 1997).

Weinberg, Steven, *Dreams of a Final Theory: The Search for the Fundamental Laws of Nature* (Hutchinson Radius, 1993).

Pictorial

Abbott, R. Tucker, *Seashells of the World* (Golden Press, 1985).

Weidensaul, Scott, *Fossil Identifier* (Quintet, 1992).

Wolfe, Art and Barbara Sleeper, *Wild Cats of the World* (Crown, 1995).

Quantum mechanics

Feynman, Richard P., *QED: The Strange Theory of Light and Matter* (Penguin Books, 1990).

Greene, Brian, *The Hidden Reality* (Knopf 2011).

Gribbin, John, *In Search of Schrödinger's Cat* (Black Swan, 1992).

Hey, Tony and Patrick Walters, *The Quantum Universe* (Cambridge University Press, 1987).

Hoffman, Banesh, *The Strange Story of the Quantum* (Pelican, 1959).

Relativity and cosmology

Chown, Marcus, *Afterglow of Creation* (Arrow Books, 1993).

Davies, Paul (ed.), *The New Physics* (Cambridge University Press, 1989).

Layzer, David, *Cosmogenesis: The Growth of Order in the Universe* (Oxford University Press, 1990).

Luminet, Jean-Pierre, *Black Holes* (Cambridge University Press, 1992).

Stewart, Ian, *Calculating the Cosmos* (Profile, 2016).

INDEX

ACKNOWLEDGMENTS

The publisher would like to thank the following for permission to reproduce copyright material:

Alamy
113: Walter Zerla

Alice O'Toole and Thomas Vetter
57L-R

Archive.org
156

Couette Taylor
91

Dover Books
Cover (Vitruvian man), 49L, 51

FLPA Images
142B: Michael & Patricia Fogden/Minden Pictures

Getty Images
33T: Bettmann, 40: Joel Sartore, 44L: Clive Streeter, 45: Antenna, 66: Alfred Pasieka, 90T: Arctic-Images, 101: Nutexzles, 103: Mark Schneider, 124: Victor R. Boswell, 133: Piotr Powietrzynski, 136: David Burder, 140: Benjamin Rondel, 143: E.M. Pasieka, 145B: Don Farrall, 149TR: Bettman, 164: Paul Starosta, 180: Alexander Safonov, 186: Jonathan Blair, 211: Anne Maenurm

iStock
111: Daniel Stein, 173

Library of Congress, Washington DC
19, 148L, 148R, 149TL

NASA
Cover (Jupiter), 2, 8, 16, 17T, 17B, 26, 27, 28, 31T, 31B, 35L, 35R, 37, 47, 70L, 70R, 71L, 71R, 106, 128, 139, 146, 159TL, 159TR, 159BL, 159BR, 168TL, 171B, 178, 185, 187L, 189L, 195, 198, 201T

Nature Picture Library
43B: Ingo Arndt, 142T: Ingo Arndt

M. C. Escher
79B: M. C. Escher's "Circle Limit IV" © 2016 The M.C. Escher Company, The Netherlands. All rights reserved. (www.mcescher.com)

Mike Field
165 L & R: Created by Mike Field using ideas described in *Symmetry of Chaos* by Mike Field and Marty Golubitsky, (Oxford University Press, 1992).

Science Photo Library
Cover (nautilus), cover (starfish): D. Roberts, 3L-R: Ted Kinsman, 5C: Wim Van Egmond, 11: Pekka Parviainen, 21L: Eye of Science, 23, 33B: Gustoimages, 38: Goronwy Tudor Jones, University of Birmingham, 43T: Power and Syred, 49R: Bert Myers, 53R: Alfred Pasieka, 54: Goronwy Tudor Jones, University of Birmingham, 56: Alfred Pasieka, 59: Russell Kightley, 61: David Parker, 62: Wim Van Egmond, 65T: D. Roberts, 68: Gustoimages, 72T, 77T: Steve Gschmeissner, 95L-R: Ted Kinsman, 122, 138: Massimo Brega, 145T: Mehau Kulyk, 150T: Edward Kinsman, 150B: Dr Jeremy Burgess, 151: Alfred Pasieka, 168TR, 176R: Scott Camazine, 187R: Geological Survey of Canada, 189R: Julian Baum, 193: Mark Garlick

Shutterstock
Cover (bubbles and snowflakes), 5T, 5B, 9L, 12, 13, 14, 15L, 15R, 18, 20, 21R, 24L, 24R, 25L, 25R, 29, 34, 58, 67, 69, 76, 77B, 79T, 81, 86L, 86R, 87L, 87R, 88, 89L, 89R, 90B, 92, 96L, 104, 108, 110, 117, 120L, 120R, 127, 129, 130-131BL-BR, 131, 135, 141, 154L, 154R, 155, 163, 167, 168TC, 181B, 183, 191T, 191C, 191B, 207T, 208-209L-R, 210B, 214-15L-R

Dr. J. M. T. Thompson
153: Reproduced from *Instabilities and Catastrophes in Science and Engineering* by J. M. T. Thompson (John Wiley and Sons, 1982).

Wikipedia
204T: Strebe, 205: Strebe

All reasonable efforts have been made to trace copyright holders and to obtain their permission for the use of copyright material. The publisher apologizes for any omissions and will gratefully incorporate any corrections in future reprints if notified.